Timo Pfau

Real-time coherent optical receivers

Timo Pfau

Real-time coherent optical receivers

Theoretical description and real-time implementation of digital signal processing algorithms for coherent optical receivers

Südwestdeutscher Verlag für Hochschulschriften

Impressum/Imprint (nur für Deutschland/ only for Germany)
Bibliografische Information der Deutschen Nationalbibliothek: Die Deutsche Nationalbibliothek verzeichnet diese Publikation in der Deutschen Nationalbibliografie; detaillierte bibliografische Daten sind im Internet über http://dnb.d-nb.de abrufbar.
Alle in diesem Buch genannten Marken und Produktnamen unterliegen warenzeichen-, marken- oder patentrechtlichem Schutz bzw. sind Warenzeichen oder eingetragene Warenzeichen der jeweiligen Inhaber. Die Wiedergabe von Marken, Produktnamen, Gebrauchsnamen, Handelsnamen, Warenbezeichnungen u.s.w. in diesem Werk berechtigt auch ohne besondere Kennzeichnung nicht zu der Annahme, dass solche Namen im Sinne der Warenzeichen- und Markenschutzgesetzgebung als frei zu betrachten wären und daher von jedermann benutzt werden dürften.

Verlag: Südwestdeutscher Verlag für Hochschulschriften Aktiengesellschaft & Co. KG
Dudweiler Landstr. 99, 66123 Saarbrücken, Deutschland
Telefon +49 681 37 20 271-1, Telefax +49 681 37 20 271-0, Email: info@svh-verlag.de
Zugl.: Universität Paderborn, Diss., 2009

Herstellung in Deutschland:
Schaltungsdienst Lange o.H.G., Berlin
Books on Demand GmbH, Norderstedt
Reha GmbH, Saarbrücken
Amazon Distribution GmbH, Leipzig
ISBN: 978-3-8381-0594-9

Imprint (only for USA, GB)
Bibliographic information published by the Deutsche Nationalbibliothek: The Deutsche Nationalbibliothek lists this publication in the Deutsche Nationalbibliografie; detailed bibliographic data are available in the Internet at http://dnb.d-nb.de.
Any brand names and product names mentioned in this book are subject to trademark, brand or patent protection and are trademarks or registered trademarks of their respective holders. The use of brand names, product names, common names, trade names, product descriptions etc. even without a particular marking in this works is in no way to be construed to mean that such names may be regarded as unrestricted in respect of trademark and brand protection legislation and could thus be used by anyone.

Publisher:
Südwestdeutscher Verlag für Hochschulschriften Aktiengesellschaft & Co. KG
Dudweiler Landstr. 99, 66123 Saarbrücken, Germany
Phone +49 681 37 20 271-1, Fax +49 681 37 20 271-0, Email: info@svh-verlag.de

Copyright © 2009 by the author and Südwestdeutscher Verlag für Hochschulschriften Aktiengesellschaft & Co. KG and licensors
All rights reserved. Saarbrücken 2009

Printed in the U.S.A.
Printed in the U.K. by (see last page)
ISBN: 978-3-8381-0594-9

Abstract

The continuous increase of the worldwide data traffic demands new concepts for data transmission in the optical fiber-based backbone networks. One promising way to increase the capacity of the existing fiber infrastructure is to use multilevel modulation formats in combination with polarization-multiplexing and coherent detection. Though elaborate transmitters and receivers are required to transmit multiple bits per symbol, but this also enables a very efficient utilization of the available bandwidth. The development of coherent optical receivers thereby profits from advancements in integrated circuit technologies that allow the digital realization of the required signal processing.

In this book all necessary algorithms for the signal processing in a coherent digital receiver are presented. The main focus thereby lies on the algorithms for polarization control and carrier recovery. A digital polarization control is required to realize a polarization-multiplexed transmission system without optical polarization control. Both a non-data-aided and a decision-directed polarization control algorithm are presented. For the latter an extension is proposed to enable also the compensation of intersymbol interference.

The most time-critical task in coherent receivers for optical transmission systems is it to recover the carrier phase from the received symbols. Due to the large linewidth of the distributed feedback (DFB) lasers employed in commercial systems a high phase noise tolerance is required. Several algorithms have been proposed to solve this problem. This book compares the different approaches at the example of the quadrature phase shift keying (QPSK) modulation format. Additionally a novel feed-forward carrier recovery for arbitrary quadrature amplitude modulation (QAM) constellations is presented. Together with the other carrier recovery schemes it is analyzed for QPSK, but additionally also for higher-level square QAM.

Finally the results of the real-time implementation of a polarization-multiplexed synchronous optical QPSK transmission system are presented, which was developed in the framework of the synQPSK project funded by the European Commission. The algorithms implemented in the coherent receiver and their parameters are optimized based on the simulation results of this thesis. Both the single-polarization QPSK transmission system and the polarization-multiplexed QPSK transmission system presented in this book are the worldwide first that were realized with a real-time coherent digital receiver and standard DFB lasers.

Zusammenfassung

Der kontinuierliche Anstieg des weltweiten Datenverkehrs erfordert neue Datenübertragungskonzepte für die auf optischen Glasfasern basierenden Backbone-Netze. Eine vielversprechende Möglichkeit, die Kapazität der bestehenden Glasfaser-Infrastruktur zu erhöhen, ist der Einsatz von mehrstufigen Modulationsverfahren in Kombination mit Polarisationsmultiplex und kohärentem Empfang. Zwar werden aufwendige Sender und Empfänger benötigt, um mehrere Bit pro Symbol zu übertragen, aber das ermöglicht auch eine sehr effiziente Nutzung der verfügbaren Bandbreite. Die Entwicklung kohärenter optischer Empfänger profitiert dabei von den Fortschritten in der integrierten Schaltungstechnik, die eine digitale Realisierung der erforderlichen Signalverarbeitung ermöglicht.

In diesem Buch werden alle zur Signalverarbeitung in einem kohärenten digitalen Empfänger benötigten Algorithmen vorgestellt. Der Schwerpunkt liegt dabei auf den Algorithmen zur Polarisationsregelung und Trägerrückgewinnung. Eine digitale Polarisationsregelung wird benötigt, um ein Übertragungssystem mit Polarisationsmultiplex ohne optische Polarisationsregelung zu realisieren. Sowohl ein datenunabhängiger und ein entscheidungsgesteuerter Polarisationsregel-Algorithmus werden vorgestellt. Für letzteren wird eine Erweiterung vorgeschlagen, die zusätzlich die Kompensation von Intersymbolstörungen ermöglicht.

Die zeitkritischste Aufgabe für den kohärenten Empfänger eines optischen Übertragungssystems ist die Rückgewinnung der Trägerphase aus den empfangenen Symbolen. Aufgrund der hohen Linienbreite der in kommerziellen Systemen eingesetzten DFB-Laser wird eine hohe Phasenrauschtoleranz benötigt. Mehrere Algorithmen wurden zur Lösung dieses Problems vorgeschlagen. Dieses Buch vergleicht die verschiedenen Ansätze am Beispiel der Quadratur-Phasenumtastung (QPSK). Zusätzlich wird eine neuartige vorwärtsgekoppelte Trägerrückgewinnung für Quadratur-Amplitudenmodulation (QAM) mit beliebigen Konstellationen vorgestellt. Zusammen mit den anderen Verfahren zur Trägerrückgewinnung wird sie für QPSK, aber auch für höherstufige quadratische QAM analysiert.

Schließlich werden die Ergebnisse einer Echtzeit-Implementierung eines synchronen optischen Übertragungssystems mit Polarisationsmultiplex vorgestellt, das im Rahmen des EU-geförderten synQPSK-Projekts entwickelt wurde. Die in dem kohärenten Empfänger implementierten Algorithmen und ihre zugehörigen Parameter wurden mithilfe der Simulationsergebnisse dieser Arbeit optimiert. Sowohl das QPSK Übertragungssystem mit einfacher Polarisation als auch das QPSK Übertragungssystem mit Polarisationsmultiplex sind weltweit die ersten, die mit einem kohärenten digitalen Echtzeit-Empfänger und Standard-DFB-Lasern realisiert wurden.

Table of contents

1 INTRODUCTION 1
  1.1 THE EUROPEAN SYNQPSK PROJECT 2
  1.2 OUTLINE OF THE BOOK 3

2 FUNDAMENTALS 5
  2.1 M-ARY QUADRATURE AMPLITUDE MODULATION 5
    2.1.1 *QAM constellations with equidistant-phases* 5
    2.1.2 *Square QAM constellations* 7
    2.1.3 *Differential encoding and decoding* 9
  2.2 COHERENT OPTICAL QAM TRANSMISSION SYSTEM 11
    2.2.1 *Optical QAM transmitter* 11
    2.2.2 *Polarization-multiplexed QAM transmitter* 12
    2.2.3 *Optical transmission link impairments* 13
    2.2.4 *Coherent optical QAM receiver with digital signal processing* 16

3 DIGITAL SIGNAL PROCESSING ALGORITHMS FOR COHERENT OPTICAL RECEIVERS 23
  3.1 CONSTRAINTS FOR ALGORITHMS IN DIGITAL RECEIVERS FOR COHERENT OPTICAL COMMUNICATION 23
    3.1.1 *Feasibility of parallel processing* 24
    3.1.2 *Hardware efficiency* 25
    3.1.3 *Tolerance against feedback delays* 26
  3.2 CLOCK RECOVERY 28
  3.3 POLARIZATION CONTROL & EQUALIZATION 29
    3.3.1 *Non-data-aided polarization control* 29
    3.3.2 *Decision-directed polarization control* 30
    3.3.3 *Decision-directed ISI compensation* 32
  3.4 FEED-FORWARD CARRIER RECOVERY 34
    3.4.1 *Viterbi & Viterbi algorithm* 34
    3.4.2 *Weighted Viterbi & Viterbi algorithm* 35
    3.4.3 *Barycenter algorithm* 36
    3.4.4 *Feed-forward carrier recovery for arbitrary QAM constellations* 40
    3.4.5 *Hardware effort* 43
  3.5 DATA RECOVERY 44
    3.5.1 *Data recovery for QAM constellations with equidistant-phases* 44
    3.5.2 *Data recovery for square QAM constellations* 45
  3.6 INTERMEDIATE FREQUENCY CONTROL 46
    3.6.1 *External LO frequency control* 46
    3.6.2 *Internal intermediate frequency compensation* 46

4 SIMULATION RESULTS 47
  4.1 QPSK CARRIER RECOVERY 47
    4.1.1 *QPSK carrier phase estimator efficiency and mean squared error* ..48
    4.1.2 *QPSK phase noise tolerance* 51
    4.1.3 *QPSK analog-to-digital converter resolution* 57
    4.1.4 *QPSK phase resolution* 58
  4.2 QAM CARRIER RECOVERY 59

Table of contents

- 4.2.1 Square QAM phase angle resolution 59
- 4.2.2 Square QAM phase estimator efficiency 60
- 4.2.3 Square QAM phase noise tolerance 66
- 4.2.4 Square QAM analog-to-digital converter resolution 69
- 4.2.5 Square QAM internal resolutions 70
- 4.3 POLARIZATION CONTROL AND PMD COMPENSATION 71
  - 4.3.1 Comparison of polarization control algorithms 71
  - 4.3.2 Verification of the ISI compensation algorithm 74

5 IMPLEMENTATION OF A SYNCHRONOUS OPTICAL QPSK TRANSMISSION SYSTEM WITH REAL-TIME COHERENT DIGITAL RECEIVER 83

- 5.1 SINGLE-POLARIZATION SYNCHRONOUS QPSK TRANSMISSION WITH REAL-TIME FPGA-BASED COHERENT RECEIVER 83
  - 5.1.1 Single-polarization synchronous QPSK transmission setup 83
  - 5.1.2 Self-homodyne experiment results at 800 Mb/s 86
  - 5.1.3 Intradyne experiment results at 800 Mb/s 87
  - 5.1.4 Intradyne experiment results at 1.6 Gb/s 88
  - 5.1.5 System optimizations & comparison of 90° hybrid with 3x3 coupler ... 89
  - 5.1.6 Comparison of experimental with simulation results 91
- 5.2 POLARIZATION-MULTIPLEXED SYNCHRONOUS QPSK TRANSMISSION WITH REAL-TIME FPGA-BASED COHERENT RECEIVER 92
  - 5.2.1 Polarization-multiplexed QPSK transmission setup 92
  - 5.2.2 Influence of different carrier recovery filter widths 98
  - 5.2.3 Polarization tracking capability 99
  - 5.2.4 Polarization tracking capability with optimized VHDL code 101
  - 5.2.5 Influence of PDL on the receiver sensitivity 102
- 5.3 POLARIZATION-MULTIPLEXED SYNCHRONOUS QPSK TRANSMISSION WITH REAL-TIME ASIC BASED COHERENT RECEIVER 103
  - 5.3.1 Transmission with and without polarization crosstalk 104
  - 5.3.2 Influence of different carrier recovery filter widths 105
  - 5.3.3 Single-polarization vs. polarization-multiplexed QPSK transmission 105

6 DISCUSSION 107

7 SUMMARY 109

8 OUTLOOK 111

9 BIBLIOGRAPHY 113

10 LIST OF FIGURES & TABLES 118

Glossary

Latin symbols

Variable	Unit	Description
$\tilde{a}$	V	Electrical drive signal of upper MZM
$\tilde{b}$	V	Electrical drive signal of lower MZM
Δs	rad	Gaussian distributed random variable for continuous phase noise
$\hat{\varphi}$	rad	Estimated carrier phase
Δf	Hz	Sum laser linewidth
Δf_{3dB}	Hz	Full width at half maximum
Δf_{DFB}	Hz	DFB laser linewidth
Δf_{ECL}	Hz	ECL linewidth
B		Number of test carrier phase angles
b_i		i-th input bit sequence into the transmitter
b_{min}		Index of minimum squared distance sum
B_r	Hz	Reference bandwidth
$\underline{c}$		Constellation point in the complex plane
c	m/s	Light velocity
c_t	s	Control time constant
$\underline{c}_k$		Transmitted complex symbol
D_{CD}	s/m²	Chromatic dispersion parameter
d_i		Distance of test sample to closest constellation point in i-th block
D_{PMD}	s/$\sqrt{\text{m}}$	Polarization mode dispersion parameter
$e(N_{CR})$		Estimator efficiency for filter half width N_{CR}
$\underline{E}_a$	V/m	Output electrical field of the upper MZM
$\underline{E}_b$	V/m	Output electrical field of the lower MZM
e_{CLK}		Clock phase error signal
$\underline{E}_{CW}$	V/m	Electrical field of the transmitter laser
$\underline{E}_i$	V/m	Electrical field of i-th optical 90° hybrid output
$\underline{E}_l$	V/m	Input electrical field into the lower MZM
$\underline{E}_{LO}$	V/m	Electrical field of local oscillator signal
$\underline{E}_{RX}$	V/m	Input electrical field of the optical receiver
E_S	J	Energy per symbol
$\underline{E}_{TX}$	V/m	Output electrical field of the optical transmitter
$\underline{E}_u$	V/m	Input electrical field into the upper MZM
F		Differential coding penalty
f_c	Hz	Carrier frequency
g		Control gain

Glossary

I_I	A	Differential output current of inphase photodiodes
I_Q	A	Differential output current of quadrature photodiodes
J		Fiber Jones matrix
k		Discrete time index
K	V/A	Transimpedance amplifier transfer ratio
l		Number of pipeline stages
L_{fiber}	m	Fiber length
L_{PMDE}		PMD emulator filter length
M		QAM modulation level (number of constellation points)
m		Number of parallel modules
M		Polarization control matrix
$\mathbf{M}_i$		Dispersion compensation matrix of i-th tap
n		Refractive index
$\underline{n}$		Complex Gaussian noise variable
N_0	W/Hz	Noise power spectral density
n_a		Amplitude number
N_{CR}		Carrier recovery filter half width
n_d		Differential half-plane/quadrant/sector number
n_i		Inphase number
n_j		Jump number
N_{PMDC}		PMD compensator filter half width
n_q		Quadrature number
n_t		Transmitter half-plane/quadrant/sector number
P	W	Optical power
p		Number of sectors for equidistant-phase constellations
P_{in}	W	Fiber input power
P_{LO}	W	Local oscillator power
P_N	W	Optical noise power
P_{out}	W	Fiber output power
P_S	W	Optical signal power
Q		Correlation matrix for polarization control
$\mathbf{Q}_i$		Correlation matrix for i-th tap for dispersion compensation
R	A/W	Photodiode responsivity
R_b	b/s	Bit rate
R_S	baud	Symbol/Baud rate
s_i		Squared distance sum in i-th block
t	s	Time
C		Optical 3 dB coupler transfer matrix
T		Polarization control error matrix for CMA
T_b	s	Bit duration
T_S	s	Symbol duration
u		Power parameter for Viterbi & Viterbi carrier recovery
$\underline{U}$		Carrier recovery filter input

U_I	V	Transimpedance amplifier output signal (inphase)
U_Q	V	Transimpedance amplifier output signal (quadrature)
$\underline{V}$		Carrier recovery filter output
v_g	m/s	Group velocity
v_i		i-th Wiener filter coefficient
W		Number of averaged correlation matrices
$\underline{X}$		Discrete signal after carrier recovery
x		FIR/IIR filter input signal
$\underline{Y}$		Discrete signal after polarization control and dispersion compensation
y		FIR/IIR filter output signal
$\underline{Z}$		Discrete signal after analog-to-digital converter
z_i		i-th output signal of 3x3 coupler

Greek symbols

Variable	Unit	Description
σ_n^2		Gaussian noise variance
ϑ	rad	Modulation free symbol phase
ϕ	rad	(S)MLPA filter cell output
γ	rad	Symmetrie angle of constellation diagram
α		Fiber attenuation coefficient
α_i		i-th FIR tap coefficient
α_{PDL}		PDL coefficient
β		Propagation constant
β_i		i-th IIR tap coefficient
δ, ε	rad	Phase offset parameters of Jones matrix
Δ		Processing delay
$\Delta\tau_{DGD}$	s	Differential group delay
$\Delta\psi$	rad	Gaussian distributed random variable for discrete phase noise
θ, ζ	rad	(S)MLPA filter cell inputs
λ	m	Wavelength
υ	rad	Polarization cross-talk parameter of Jones matrix
φ_{CLK}	rad	Clock phase
φ_i	rad	Test carrier phase of i-th block
χ_i		Correlation factor between 0-th and i-th dispersion compensation filter input
ψ_{IF}	rad	Carrier phase
ψ_{LO}	rad	Local oscillator phase
ψ_S	rad	Signal phase

ω_{IF}	Hz	Angular carrier frequency
ω_{LO}	Hz	Angular local oscillator frequency
ω_S	Hz	Angular signal frequency

Acronyms and Abbreviations

Abbreviation	Description
100GbE	100 Gigabit Ethernet
ADC	Analog-to-Digital Converter
ASE	Amplified Spontaneous Emission
ASIC	Application-specific Integrated Circuit
ASK	Amplitude Shift Keying
AWG	Arrayed-waveguide Grating
AWGN	Additive White Gaussian Noise
BER	Bit Error Rate
BERT	Bit Error Rate Tester
Bit	Binary digit
BPF	Bandpass Filter
BPSK	Binary Phase Shift Keying
CD	Chromatic Dispersion
CMA	Constant Modulus Algorithm
CMOS	Complementary Metal–Oxide–Semiconductor
CORDIC	Coordinate Rotation Digital Computer
CRLB	Cramér-Rao Lower Bound
CW	Continuous Wave
DAC	Digital-to-Analog Converter
DBPSK	Differential Binary Phase Shift Keying
DCF	Dispersion Compensating Fiber
DD	Decision-Directed
DEMUX	Demultiplexer
DFB	Distributed Feedback
DGD	Differential Group Delay
DQPSK	Differential Quadrature Phase Shift Keying
DSPU	Digital Signal Processing Unit
DWDM	Dense Wavelength Division Multiplexing
ECL	External-cavity Laser
ECOC	European Conference on Optical Communication
EDFA	Erbium-Doped Fiber Amplifier
FF	Flip-Flop
FFT	Fast Fourier Transform

FIR	Finite Impulse Response
FP6	6th Framework Programme
FPGA	Field-Programmable Gate Array
FWHM	Full Width at Half Maximum
GVD	Group Velocity Dispersion
HWP	Half-Wave Plate
IEEE	Institute of Electrical and Electronics Engineers
IF	Intermediate Frequency
IFFT	Inverse Fast Fourier Transform
IIR	Infinite Impulse Response
ISI	Intersymbol Interference
ITU-T	International Telecommunication Union - Telecommunication Standardization Sector
LO	Local Oscillator
LUT	Loop-Up Table
MGT	Multi-Gigabit Transceiver
MLPA	Maximum Likelihood Phase Approximation
M-QAM	M-ary Quadrature Amplitude Modulation
MSE	Mean Squared Error
MUX	Multiplexer
MZM	Mach-Zehnder-Modulator
NDA	Non-Data-Aided
NFOEC	National Fiber Optic Engineers Conference
OFC	Optical Fiber Communication Conference and Exposition
OFDM	Orthogonal Frequency Division Multiplexing
ONT	Optische Nachtichtentechnik und Hochfrequenztechnik (Optical Communication and High Frequency Engineering)
OOK	On-Off-Keying
OSNR	Optical Signal-to-Noise Ration
PBC	Polarization beam combiner
PBS	Polarization beam splitter
PC	Personal Computer
PDG	Polarization-dependent Gain
PDL	Polarization-dependent Loss
PLL	Phase-locked Loop
PMD	Polarization Mode Dispersion
PMDC	Polarization Mode Dispersion Compensator
PMDE	Polarization Mode Dispersion Emulator
PM-QPSK	Polarization-Multiplexed Quadrature Phase Shift Keying
PRBS	Pseudo-Random Binary Sequence
PSK	Phase Shift Keying
QAM	Quadrature Amplitude Modulation
QPSK	Quadrature Phase Shift Keying
QWP	Quarter-Wave Plate

Glossary

RTL	Register Transfer Level
SCT	Schaltungstechnik (System and Circuit Technology)
SMF	Single-Mode fiber
SMLPA	Selective Maximum Likelihood Phase Approximation
SNR	Signal to Noise Ratio
SOP	State of Polarization
SPM	Self-Phase Modulation
V&V	Viterbi & Viterbi
VCO	Voltage-Controlled Oscillator
VHDL	VHSIC Hardware Description Language
VHSIC	Very High Speed Integrated Circuit
VOA	Variable Optical Attenuator
WDM	Wavelength Division Multiplexing
XPM	Cross-Phase Modulation

1 Introduction

Coherent optical receivers that use either homodyne or heterodyne detection have significant advantages over traditional optical direct detection receivers because they linearly down-convert the optical signal to electrical signals. Therefore the receiver sensitivity is shot-noise limited, if the local oscillator (LO) power is sufficiently high.

In the 1980s this property of high receiver sensitivity directed a lot of research towards the development and implementation of coherent long-distance optical transmission systems without repeaters [1; 2; 3; 4]. But the invention of the erbium-doped fiber amplifier (EDFA) and its fast deployment in commercial transmission systems dramatically reduced the interest in coherent technologies [5; 6].

In EDFA-based systems amplified spontaneous emission (ASE) rather than shot noise determines the signal-to-noise ratio (SNR), which made the shot-noise limited receiver sensitivity of coherent receivers less significant. Additional technical difficulties inherent in coherent receivers also prevented further investigations. The disadvantage of heterodyne receivers is that an intermediate frequency (IF) higher than the symbol rate is required. Thus the receiver bandwidth must be more than twice as large as for baseband and direct detection receivers. The homodyne receiver operates at the baseband, but requires a stable locking of the transmitter and local oscillator frequency and phase. With standard distributed feedback (DFB) lasers stable locking using a phase-locked loop (PLL) could not be demonstrated [7]. Coherent receivers with analog feed-forward carrier recovery showed sufficient phase noise tolerance [8], but could not prevail over the less complex direct detection receivers.

In contrast the EDFA technology revolutionized research in optical communication in the 1990s. Thanks to the large bandwidth of EDFAs wavelength division multiplexing (WDM) techniques became possible and dramatically increased the transmission capacity of optical fibers.

In recent years research about coherent receivers experienced a revival [9]. Due to the ever-increasing bandwidth demand researchers are looking for ways to exploit the optical bandwidth more efficiently by using coherent transmission with multilevel modulation formats. The development thereby profited from the fact that over the past years the bandwidth and clock frequencies for digital signal processing circuits increased faster than the symbol rate for optical communication. Therefore the electrical signals in a coherent receiver can now be processed in a digital signal processing unit (DSPU). By means of feed-forward carrier recovery the inphase and quadrature component of the complex amplitude of the optical carrier is recovered digitally and in a stable manner [10; 11]. Moreover, all linear optical distortions (polarization transformations, polarization mode dispersion, chromatic dispersion) can theoretically be equalized without any losses [12; 13].

1 Introduction

The main research focus was laid on the investigation of synchronous optical quadrature phase shift keying (QPSK) combined with polarization division multiplex. Compared to standard on-off-keying (OOK) the line rate is 4 times lower, the needed number of photons per bit less than half as high, the tolerance to chromatic dispersion about 5 times better, the tolerance to polarization mode dispersion about 3 times better, and the tolerance against fiber nonlinearities, in particular cross phase modulation, is excellent [14]. Therefore it is an extremely attractive modulation format for metropolitan-area and long-haul fiber communication. Distinct advantages exist also over other modulation formats, such as duobinary modulation, differential binary phase shift keying (DBPSK) or differential quadrature phase shift keying (DQPSK) [9; 15].

The first transmission experiments with coherent digital receivers were realized using digital storage oscilloscopes and offline signal processing in a personal computer (PC) [16]. The reason was that some key components to realize a real-time coherent digital receiver did not exist yet. For this reason in 2004 the University of Paderborn, Photline Technologies, CeLight Israel and the Innovative Processing AG started the synQPSK-project, which aimed at the development these key components.

But not only QPSK attracts the attention of the research community, also higher level quadrature amplitude modulation (QAM) with the main focus on square QAM constellations is interesting as it allows to increase the spectral efficiency even beyond the one of polarization-multiplexed QPSK [17; 18]. Although high-level QAM is more susceptible to noise, which makes it less attractive for long-haul applications, but its ultimate spectral efficiency makes QAM very interesting for metropolitan and regional area networks, especially for next-generation networks beyond 100 Gb/s.

But as for coherent QPSK transmission the key components were missing in 2004, today even the main key algorithm for coherent QAM transmission with high-level constellations is not available: A feed-forward carrier recovery algorithm with a sufficiently high phase noise tolerance that allows the employment of standard DFB lasers.

1.1 The European synQPSK project

The synQPSK project, funded by the European Commission within the 6th Framework Programme (FP6) under the contract 004631, was started on July 1, 2004. The research consortium was coordinated by the University of Paderborn from Germany with the working groups ONT (Optical Communication and High Frequency Engineering) and SCT (System and Circuit Technology). The additional partners were Photline Technologies from France, CeLight Israel and the Innovative Processing AG from Germany, which latter was replaced after the first project year by the University of Duisburg-Essen, Germany. The overall project goal was to develop all necessary components that could not be found on the market for a synchronous optical QPSK transmission system combined with polarization division multiplex, and to validate them in a 10 Gbaud, 40 Gb/s "synQPSK" testbed.

The identified key components were LiNbO₃ QPSK modulators required in the transmitter, integrated coherent receiver frontends consisting of LiNbO₃ optical 90° hybrids co-packaged with InP balanced photoreceivers, and SiGe/CMOS integrated electronic circuits for analog-to-digital conversion and digital signal processing. Figure 1.1 shows the layout of the synQPSK project and links the consortium partners to their respective development task.

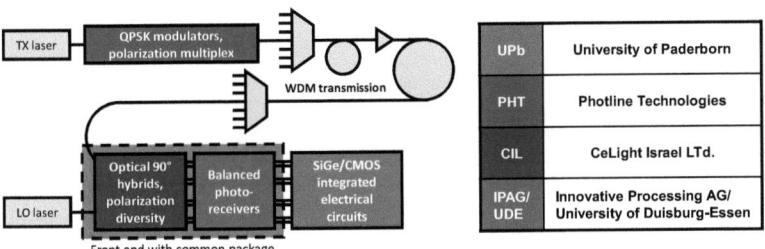

Figure 1.1: Simplified system schematic for the synQPSK project with partners' contributions highlighted

Within the University of Paderborn the development tasks were distributed as follows: The working group "Optical Communication and High Frequency Engineering" (ONT) headed by Prof. Dr.-Ing. Reinhold Noé was responsible for algorithm development, system simulations, development of high-speed analog-to-digital converters in SiGe technology and the design of full-custom demultiplexers in CMOS. Additionally the working group was responsible for the synQPSK testbed, i.e. the initial operation and validation of all components and the assembly of a fully functional coherent polarization-multiplexed QPSK transmission system.

The working group "System and Circuit Technology" (SCT) of Prof. Dr.-Ing. Ulrich Rückert was responsible for the hardware implementation of the algorithms provided by ONT, the integration of full custom demultiplexers in the DSPU standard cell design and the backend development of the CMOS application-specific integrated circuits (ASIC).

Most of the work that is related to synchronous QPSK transmission and presented in this book was conducted in the framework of the synQPSK project.

1.2 Outline of the book

At first the theoretical description of a fiber-optic transmission system with coherent receiver and digital signal processing is presented. Starting from the specification of two main classes of constellations for M-ary quadrature amplitude modulation (M-QAM), i.e. QAM constellations with equidistant-phases and square QAM constellations, and their generation in an optical transmitter is described. Then the main distortions that occur while the optical signal is traveling though the fiber are summarized, and finally the coherent detection of the signal in an optical polarization diversity receiver with subsequent analog-to-digital conversion is explained.

1 Introduction

Before going into detail in chapter 3 about the algorithms required in a digital signal processing unit (DSPU) of a coherent optical receiver, chapter 3.1 summarizes the constraints for these algorithms to be suitable for real-time implementation. Then the algorithms for clock recovery, polarization control, dispersion compensation, carrier recovery and intermediate frequency control are described. Two of the core elements of this book are presented in this chapter: The dispersion compensation algorithm as well as the carrier recovery for arbitrary QAM constellations were developed by the author.

In chapter 4 the simulation results for polarization control, dispersion compensation and carrier recovery are presented. The purpose of the simulations is to demonstrate the applicability and performance of the newly proposed algorithms, and for the QPSK carrier recovery to compare the performance against state-of-the-art techniques. Additionally the simulations were required to determine the key parameters for a hardware implementation of a real-time synchronous QPSK receiver in the framework of the synQPSK project.

The setup for this hardware implementation and the measurement results derived from it are finally outlined in chapter 5. The structure of this chapter follows the implementation sequence of the system, from single-polarization QPSK transmission to polarization-multiplexed QPSK transmission, both based on a field-programmable gate array (FPGA) for digital signal processing, to the final polarization-multiplexed synchronous QPSK setup with specifically developed SiGe and CMOS application-specific integrated circuits (ASIC). A discussion of the achieved results followed by a summary and an outlook close the book.

2 Fundamentals

In digital communication systems information is sent from a source through a transmission channel to a remote sink. In fiber-optic communication the source is represented by the optical transmitter. According to the applied modulation format it maps the transmitted sequence of binary digits (bit) with the bit rate $R_b = 1/T_b$ to symbols with the symbol or baud rate $R_S = 1/T_S$. T_b and T_S are the bit and symbol duration, respectively. The ratio R_b/R_S specifies the number of bits per symbol and is a measure for the spectral efficiency of the modulation format. The symbols are impressed on a carrier signal that can be sent through the optical fiber to the optical receiver. The receiver then recovers the symbols from the received signal and reconstructs the bit sequence. Although there are different types of optical receivers, this book only considers coherent optical receivers. In the following these different components of a fiber-optic transmission system are described in more detail.

2.1 M-ary quadrature amplitude modulation

In quadrature amplitude modulation (QAM) data is transported by modulating the amplitude of two carriers, which have the same frequency f_c but are 90° out of phase. They can therefore be called quadrature carriers – hence the name of the scheme [19]. A convenient way to represent digital QAM schemes is the constellation diagram. Inphase and quadrature modulation are represented as real and imaginary parts of a complex number. The number of symbols M in the constellation diagram defines the order of a digital QAM format, which can therefore be named M-ary QAM or M-QAM.

But to specify the order of a QAM constellation is not sufficient to uniquely qualify a M-QAM format, because the M symbols can be arbitrarily distributed over the complex plane. Thus also the shape of the QAM constellation must be considered. In this book I will concentrate on the two most important kinds of shapes for QAM constellations, which are mostly used in commercial transmission systems: Equidistant-phase constellations and square QAM constellations.

2.1.1 QAM constellations with equidistant-phases

A QAM constellation scheme with equidistant-phases is also referred to as phase shift keying (PSK), if the amplitude is constant, or combined amplitude and phase shift keying (ASK-PSK), if also the amplitude is modulated. The most commonly used modulation schemes of this QAM sub-class are binary phase shift keying (BPSK) with a spectral efficiency of 1 bit/symbol and quadrature phase shift keying (QPSK) with a spectral efficiency of 2 bit/symbol. The constellation diagrams of the two schemes with the corresponding Gray-coded bit assignments are depicted in Figure 2.1. The colored areas represent the tolerable corruption by noise while the correspondent symbol is still detected correctly at the receiver.

2 Fundamentals

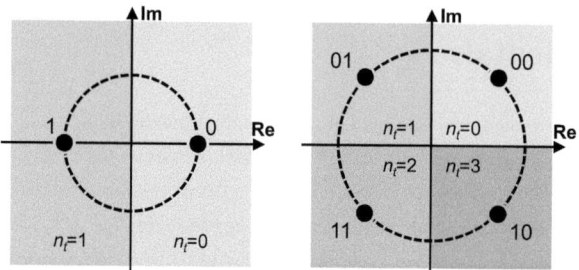

Figure 2.1: BPSK (left) and QPSK (right) constellation diagrams

The symbols positions in the complex plane for BPSK are given by the formula

$$\underline{c}_{\text{BPSK}} = \exp\{j\pi \cdot n_t\} \quad n_t \in \{0,1\} \tag{2.1}$$

and for QPSK by

$$\underline{c}_{\text{QPSK}} = \sqrt{2}\exp\left\{j\frac{\pi}{2}n_t + \frac{\pi}{4}\right\} \quad n_t \in \{0,1,2,3\}. \tag{2.2}$$

In case of BPSK n_t can be regarded as a half-plane number, in case of QPSK as a quadrant number. The bit-to-symbol assignment is calculated by converting the binary value of n_t to Gray-code [20].

Equation (2.1) and (2.2) imply that for QPSK twice the signal power is required compared to BPSK to achieve the same distance between adjacent constellation points. Thus for the same signal power the distance between adjacent symbols is reduced. This shows that increasing the order of QAM allows the transmission of more bits per symbol, but at the price of a less reliable detection at the receiver.

But also higher level modulation formats are possible. Figure 2.2 shows a ASK-8-PSK constellation diagram for a spectral efficiency of 4 bit/symbol. The symbol positions in the complex plane are given by

$$\underline{c}_{\text{ASK-8-PSK}} = n_a \exp\left\{j\frac{\pi}{4}n_t\right\} \quad \begin{aligned} n_a &\in \{1,2\} \\ n_t &\in \{0,1,...,7\} \end{aligned}. \tag{2.3}$$

The bit-to-symbol assignment depends now on n_t, which can be considered as a segment number and determines the first three bits of a symbol. The amplitude number n_a determines if the symbol is lying on the inner or outer circle represented by the last bit.

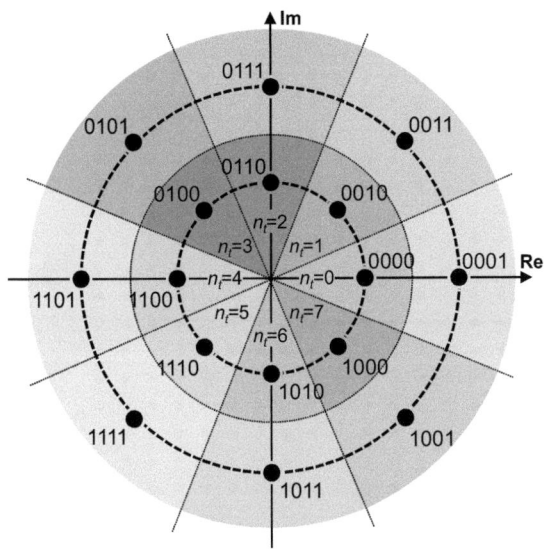

Figure 2.2: ASK-8-PSK constellation diagram

In Figure 2.2 the disadvantage of QAM constellation scheme with equidistant-phases for higher-order constellations becomes obvious. The distances to adjacent symbols are smaller for symbols on the inner circle than for the symbols on the outer circle. In systems where phase noise is dominant, this does not matter, but for systems where additive white Gaussian noise (AWGN) dominates, square QAM constellations are more tolerant against noise than equidistant-phase constellations [21].

2.1.2 Square QAM constellations

In square QAM constellations the symbols are placed on a square grid with equal vertical and horizontal spacing. Due to the uniform distribution square QAM constellations are less susceptible to AWGN than QAM constellation scheme with equidistant-phases. Figure 2.3 shows different square QAM constellation diagrams ranging from 4-QAM, which is equivalent to QPSK and has a spectral efficiency of 2 bit/symbol, to 256-QAM with a spectral efficiency of 8 bit/symbol.

2 Fundamentals

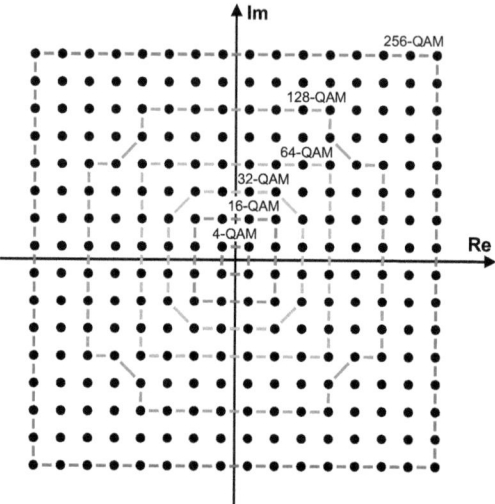

Figure 2.3: Square QAM constellation diagrams

As QPSK is the simplest square QAM, it is straightforward to describe the positions of the symbols for M-QAM by extending equation (2.2) by two new variables n_i and n_q, which describe the additional amplitude modulation along the real (inphase) and imaginary (quadrature) axis, respectively. If $\log_2(M)$ is an even number, then the number of amplitude levels on the real and imaginary axis is $\sqrt{M}$ and the positions of constellation points are given by

$$\underline{c}_{M\text{-QAM}} = \underline{c}_{\text{QPSK}} + 2\,\text{sgn}\!\left[\text{Re}\{\underline{c}_{\text{QPSK}}\}\right]\!n_i + j \cdot 2\,\text{sgn}\!\left[\text{Im}\{\underline{c}_{\text{QPSK}}\}\right]\!n_q \quad \begin{matrix} n_i \in \{0,1,...,\sqrt{M}/2-1\} \\ n_q \in \{0,1,...,\sqrt{M}/2-1\} \end{matrix}. \quad (2.4)$$

As for QPSK n_t can be considered as a quadrant number represented by two bits, and n_i and n_q each represent half of the remaining bits. It is sufficient to separately Gray-encode n_t, n_i and n_q. The resulting constellation will also be Gray-encoded.

If $\log_2(M)$ is an odd number, then the constellation diagram is not an ideal square as can be seen in Figure 2.3. But it can be easily constructed by extending the constellation diagram of the $M/2$ square QAM by adding $\sqrt{M/8}$ additional amplitude levels on the real and imaginary axes. In this case the constellation points with simultaneous $n_i \geq \sqrt{M/8}$ and $n_q \geq \sqrt{M/8}$ are unused.

The bit-to-symbol assignment for square QAM constellations is exemplified for 16-QAM in Figure 2.4. n_t is represented by the first two bits, n_i corresponds to the 3rd bit, n_q to the 4th bit. The colored areas show the tolerable corruption by noise while the corresponding symbol is still detected correctly at the receiver.

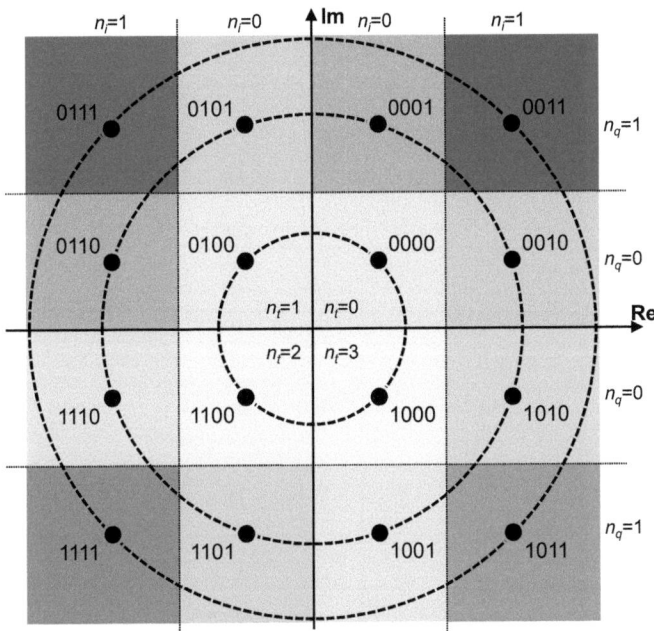

Figure 2.4: Square 16-QAM constellation diagram and bit-to-symbol assignment

By comparing Figure 2.4 to Figure 2.2 it becomes obvious why square QAM constellations are preferable in AWGN-dominated transmission systems.

2.1.3 Differential encoding and decoding

A problem in QAM detection at the receiver is that the constellations are rotationally symmetric by the angle $2\pi/n_t$. Due to this n_t-fold phase ambiguity the absolute phase rotation of the constellation introduced by the transmission channel cannot be recovered by the receiver. To overcome this problem differential encoding at the transmitter and corresponding differential decoding at the receiver can be applied [21]. Differential encoding means that the information is contained in the phase difference between two consecutive symbols rather than in the absolute phase. The drawback is that if one bit is detected wrongly the differential decoding causes two consecutive bits to be wrong. Therefore it is desirable to apply differential encoding only to as few bits as possible. This is referred to as partial differential encoding.

As the phase ambiguities of all QAM constellations presented in the sections 2.1.1 and 2.1.2 only depend on n_t, it is sufficient to solely differentially encode n_t

$$n_{d,k} = (n_{d,k-1} + n_{t,k}) \mod(\max\{n_t\}+1), \qquad (2.5)$$

where k is the discrete time index and $\max\{n_t\}+1$ is the possible number of values of n_t. Thus the range of values of n_d and n_t is the same. Differential decoding at the receiver undoes the differential encoding by calculating

$$\hat{n}_{t,k} = (\hat{n}_{d,k} - \hat{n}_{d,k-1}) \mod(\max\{n_t\}+1). \tag{2.6}$$

Figure 2.5 depicts the partial differential encoding process for a square 16-QAM constellation. The encircled symbol pairs mark deviations from ideal Gray coding due to the differential encoding process.

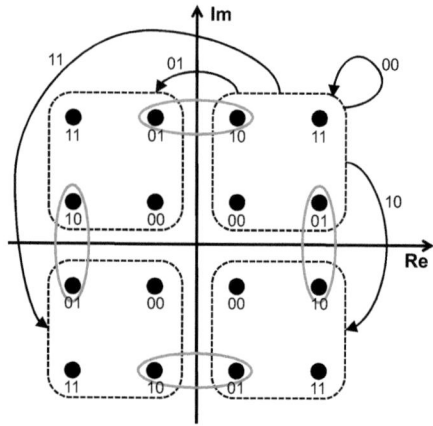

Figure 2.5: Partial differential encoding for a square 16-QAM constellation

Thus two effects degrade the performance of a transmission system if differential encoding is applied: A symbol error causes at least two bit errors due to the comparisons used in the differential decoding process described by equation (2.6), and additional errors may occur due to the deviation from ideal Gray coding. Both effects are considered in the differential coding penalty F defined as the bit error probability ratio of the differentially coded system to the non-differentially coded system [21]. In [22] it is shown that for square QAM constellations this coding penalty is given by

$$F = 1 + \frac{\log_2(M)}{2(\sqrt{M}-1)}. \tag{2.7}$$

Because the relative number of differentially encoded bits decreases as the total number of bits per symbol increases, the differential coding penalty drops from 2 for QPSK to nearly 1 for high-order QAM formats (Table 2.1).

Table 2.1: Differential coding penalty for different square QAM constellations

Constellation	Bits per symbol	Differential coding penalty F
4-QAM	2	2.00 (3.0 dB)
16-QAM	4	1.67 (2.2 dB)
64-QAM	6	1.43 (1.5 dB)
256-QAM	8	1.27 (1.0 dB)
1024-QAM	10	1.16 (0.6 dB)

An alternative to the differential encoding/decoding process is it to use framing information to resolve the phase ambiguity of the constellation diagram at the receiver [23]. But in the simulations as well as in the experiments presented in this book no framing information is transmitted. Therefore differential encoding has to be applied.

2.2 Coherent optical QAM transmission system

2.2.1 Optical QAM transmitter

There are many possible implementations for an optical QAM transmitter. In this section I present a transmitter architecture that is most commonly used and is compatible to arbitrary QAM constellations. In literature it is often referred to as IQ-modulator or nested Mach-Zehnder-modulator [24]. Figure 2.6 shows the schematic of such a transmitter.

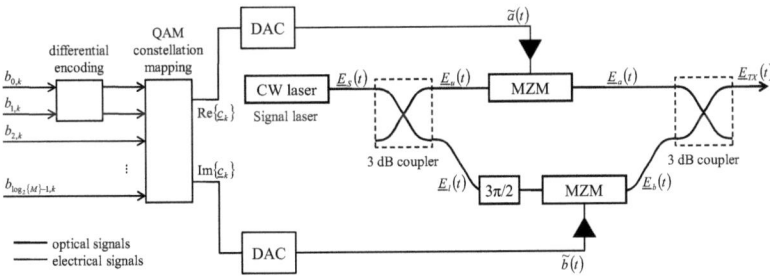

Figure 2.6: Optical QAM transmitter structure

The electrical field $\underline{E}_{CW}(t) = 2\sqrt{P_S}\, e^{j(\omega_S t + \psi_S(t))}$ generated by a continuous wave (CW) laser is split by a directional coupler with the transfer matrix

$$\mathbf{C} = \frac{1}{\sqrt{2}}\begin{bmatrix} 1 & j \\ j & 1 \end{bmatrix} \qquad (2.8)$$

2 Fundamentals

into an upper path $\underline{E}_u(t) = \sqrt{2P_S}\, e^{j(\omega_S t + \psi_S)}$ and a lower path $\underline{E}_l(t) = j\sqrt{2P_S}\, e^{j(\omega_S t + \psi_S)}$. $(1/2)|\underline{E}_{CW}|^2 = 2P_S$ is the power of the CW laser, $\omega_S/2\pi$ is the optical carrier frequency. $\underline{E}_u(t)$ in the upper path is modulated in a Mach-Zehnder modulator (MZM) by the electrical driving signal $\tilde{a}(t)$. In the lower path the electrical signal $\tilde{b}(t)$ modulates $\underline{E}_l(t)$ in a MZM. The continuous signals $\tilde{a}(t)$ and $\tilde{b}(t)$ correspond to the discrete samples $\mathrm{Re}\{\underline{c}_k\}$ and $\mathrm{Im}\{\underline{c}_k\}$, respectively. The modulated optical signals in the upper and lower paths can be written as

$$\underline{E}_a(t) = \tilde{a}(t)\sqrt{2P_S}\, e^{j(\omega_S t + \psi_S(t))}$$
$$\underline{E}_b(t) = \tilde{b}(t)\sqrt{2P_S}\, e^{j(\omega_S t + \psi_S(t))}$$
(2.9)

The additional phase shift of $3\pi/2$ in the lower path as depicted in Figure 2.6 is already considered in the equation for $\underline{E}_b(t)$. After combination in the following cross coupler we obtain the optical signal

$$\underline{E}_{TX}(t) = [a(t) + jb(t)]\sqrt{P_S}\, e^{j(\omega_S t + \psi_S)} = \underline{c}(t)\sqrt{P_S}\, e^{j(\omega_S t + \psi_S(t))}.$$
(2.10)

$\underline{E}_{TX}(t)$ is the output signal of the transmitter with the optical power P_S. At the time instants kT_S with $k = 0, \pm 1, \pm 2\ldots$ and T being the symbol duration, $\underline{c}(kT_S)$ corresponds to the discrete symbol $\underline{c}_k$ in the constellation diagram.

2.2.2 Polarization-multiplexed QAM transmitter

To generate a polarization-multiplexed transmission signal the electrical field from the CW laser must be split by a polarization beam splitter (PBS) into two branches:

$$\begin{bmatrix} \underline{E}_{CW,x}(t) \\ \underline{E}_{CW,y}(t) \end{bmatrix} = \frac{1}{\sqrt{2}} \begin{bmatrix} 2\sqrt{P_S} \\ 2\sqrt{P_S} \end{bmatrix} e^{j(\omega_S t + \psi_S(t))}$$
(2.11)

Then the signals are fed into two parallel QAM transmitter as described above. After modulation the signals are recombined in a polarization beam combiner (PBC) to form the polarization-multiplexed transmission signal

$$\begin{bmatrix} \underline{E}_{TX,x}(t) \\ \underline{E}_{TX,y}(t) \end{bmatrix} = \frac{1}{\sqrt{2}} \begin{bmatrix} \underline{c}_x(t) \\ \underline{c}_y(t) \end{bmatrix} \sqrt{P_S}\, e^{j(\omega_S t + \psi_S(t))}.$$
(2.12)

Figure 2.7 shows the structure of a polarization-multiplexed QAM transmitter.

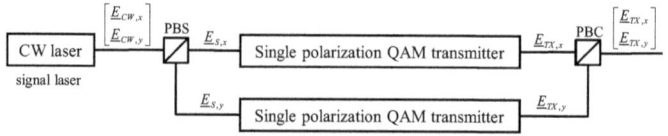

Figure 2.7: Polarization-multiplexed QAM transmitter

2.2.3 Optical transmission link impairments

This section introduces the main optical transmission link impairments that alter the signal while it travels through the fiber.

2.2.3.1 Attenuation

The attenuation caused by optical fibers limits the performance of fiber-optic communication systems by reducing the average power that reaches the receiver [24]. Since optical receivers need a certain minimum amount of power to recover the signal accurately, the transmission distance is inherently limited.

Under quite general conditions power attenuation inside an optical fiber is governed by

$$\frac{dP}{dz} = -\alpha P, \qquad (2.13)$$

where P is the optical power in the fiber. The attenuation coefficient α includes material absorption as well as other sources of power attenuation. If P_{in} is the power launched at the input of a fiber of length L_{fiber}, the output power P_{out} from (2.13) is given by

$$P_{out} = P_{in} \exp\{-\alpha L_{fiber}\}. \qquad (2.14)$$

It is customary to express α in the units of dB/km by using the relation

$$\alpha_{dB/km} = -\frac{10}{L_{fiber}} \log_{10}\left(\frac{P_{in}}{P_{out}}\right) \qquad (2.15)$$

and to refer to it as the fiber loss.

2.2.3.2 Polarization crosstalk & polarization-dependent loss

Variations in the shape of the core of a SMF cause random changes of the polarization of a pulse travelling through the fiber [24]. Therefore the state of polarization (SOP) is arbitrary at the receiver of an optical transmission system. It is common to describe the change of the SOP by a unitary Jones matrix [25] given by

$$\mathbf{J}(t) = \begin{bmatrix} \cos\{\upsilon(t)\}e^{j\delta(t)/2} & -\sin\{\upsilon(t)\}e^{-j\varepsilon(t)/2} \\ \sin\{\upsilon(t)\}e^{j\varepsilon(t)/2} & \cos\{\upsilon(t)\}e^{-j\delta(t)/2} \end{bmatrix}. \qquad (2.16)$$

The time-variant parameter $\upsilon(t)$ describes the cross-talk between the two polarization modes, $\delta(t)$ and $\varepsilon(t)$ denote the phase differences. The input signal to the receiver is then given by the fiber input signal at the transmitter multiplied by the fiber Jones matrix.

$$\begin{bmatrix} \underline{E}_{RX,x}(t) \\ \underline{E}_{RX,y}(t) \end{bmatrix} = \mathbf{J}(t) \begin{bmatrix} \underline{E}_{TX,x}(t) \\ \underline{E}_{TX,y}(t) \end{bmatrix} = \begin{bmatrix} \cos\{\upsilon(t)\}e^{j\delta(t)/2}\underline{E}_{TX,x}(t) - \sin\{\upsilon(t)\}e^{-j\varepsilon(t)/2}\underline{E}_{TX,y}(t) \\ \sin\{\upsilon(t)\}e^{j\varepsilon(t)/2}\underline{E}_{TX,x}(t) + \cos\{\upsilon(t)\}e^{-j\delta(t)/2}\underline{E}_{TX,y}(t) \end{bmatrix} \qquad (2.17)$$

The Jones matrix $\mathbf{J}(t)$ is time-variant. Slow variations of $\mathbf{J}(t)$ are caused by temperature drifts, but also very fast polarization change speeds with several krad/s on the Poincaré sphere are possible. These fast fluctuations are caused by movements of the fiber, e.g. by vibrations of DCF coils [26].

But not only the SOP changes while the signal is travelling through the fiber, the two polarization modes can also suffer from different rates of loss due to asymmetries of the fiber [24]. This effect is referred to as polarization-dependent loss (PDL). In an optical transmission system it can be modeled by a lumped PDL element placed between 2 unitary Jones matrices.

$$\begin{bmatrix} E_{RX,x}(t) \\ E_{RX,y}(t) \end{bmatrix} = \mathbf{J}_1(t) \begin{pmatrix} 1 & 0 \\ 0 & \alpha_{PDL} \end{pmatrix} \mathbf{J}_0(t) \begin{bmatrix} E_{TX,x}(t) \\ E_{TX,y}(t) \end{bmatrix} \quad (2.18)$$

It is customary to express PDL in the unit of dB by using the relation

$$\alpha_{PDL,dB} = 20 \log_{10}(\alpha_{PDL}). \quad (2.19)$$

It is also possible that $\alpha_{PDL} > 1$. In this case the effect is referred to as polarization-dependent gain (PDG).

2.2.3.3 Chromatic dispersion

Dispersion is a major source of signal distortion in optical fiber transmission systems [24; 27]. Single-mode fibers (SMF) have the advantage that intermodal dispersion is absent because the energy of the injected pulse is transported only by a Single-mode. However pulse broadening does not disappear altogether due to chromatic dispersion.

Chromatic dispersion occurs because all optical signals have a finite spectral width, and different spectral components travel with different speeds through the fiber. One cause of this velocity difference is that the refractive index $n(\omega)$ of a SMF is frequency-dependent. This is called material dispersion and it is the dominant source of chromatic dispersion in single-mode fibers. Another cause of dispersion is that the cross-sectional distribution of light within the fiber also changes for different wavelengths. Shorter wavelengths are more completely confined to the fiber core, while a larger portion of the optical power at longer wavelengths propagates in the cladding. Since the index of the core is greater than the index of the cladding, this difference in spatial distribution causes a change in propagation velocity. This phenomenon is known as waveguide dispersion. Waveguide dispersion is relatively small compared to material dispersion.

The chromatic dispersion property of an optical fiber is given by the group-velocity dispersion parameter D_{CD}, which is usually expressed in ps/nm/km [24]. In general, for a signal with an angular frequency $\omega(\beta)$ at a propagation constant β, i.e. the electromagnetic fields in the propagation direction z oscillate proportional to $e^{j(\beta z - \omega t)}$, the dispersion parameter D_{CD} is defined as

$$D_{CD} = -2\pi \frac{c}{\lambda^2} \frac{d^2\beta}{d\omega^2} = 2\pi \frac{c}{v_g^2 \lambda^2} \frac{dv_g}{d\omega}. \quad (2.20)$$

where $\lambda = 2\pi c/\omega$ is the vacuum wavelength and $v_g = d\omega/d\beta$ is the group velocity.

2.2.3.4 Polarization mode dispersion

In realistic fibers random imperfections in the circular symmetry cause the two polarizations within the fiber to travel at different speeds [24]. This phenomenon causes pulse broadening and is called polarization mode dispersion. Due to the random characteristic of the fiber imperfections the pulse broadening effect corresponds to a random walk. Thus the differential group delay (DGD) $\Delta\tau_{DGD}$ is proportional to the square root of the fiber length L_{fiber}.

$$\Delta\tau_{DGD} = D_{PMD}\sqrt{L_{fiber}} \qquad (2.21)$$

The PMD parameter D_{PMD} of the fiber is usually expressed in $ps/\sqrt{km}$ and is a measure for the asymmetry of the fiber. For a standard single-mode fiber (SMF) [ITU-T G.652] the PMD parameter is $D_{PMD} = 0.1\ ps/\sqrt{km}$ [28].

A good model to emulate PMD is to use the filter structure depicted in Figure 2.8 [29].

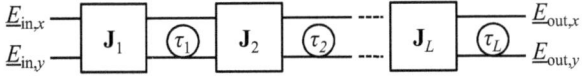

Figure 2.8: Polarization mode dispersion emulator (PMDE)

The PMD emulator (PMDE) is given by a cascade, which consists alternately of a unitary Jones matrix as described in equation (2.16) and a component that adds the additional delay τ to one of the two polarizations. Its total differential group delay depends on the values of the Jones matrices. With random Jones matrices and $\tau_1 = \tau_2 = ... = \tau_{L_{PMDE}} = \tau_{PMD}$ the expectation of $\Delta\tau_{DGD}$ becomes

$$\langle\Delta\tau_{DGD}\rangle = \tau_{PMD}\sqrt{L_{PMDE}}\ . \qquad (2.22)$$

2.2.3.5 Amplified spontaneous emission

To compensate for the fiber loss introduced in section 2.2.3.1 optical amplifiers are used to regenerate the signal before detection at the receiver. The most common amplifier is the erbium-doped fiber amplifier (EDFA) [5; 6]. It has a huge amplification window that can cover both the optical C-band (1525 nm $\leq \lambda \leq$ 1565 nm) and L-band (1570 nm $\leq \lambda \leq$ 1610 nm). The signal is amplified by being multiplexed in the doped fiber with a pump laser signal at a wavelength of 980 nm or 1480 nm. The pump laser excites the trivalent Erbium ions (Er^{+3}) into a higher energy level. By interaction with a photon at the signal wavelength the ion can decay back to a lower energy level by emitting a photon with the same wavelength as the signal. This effect is called stimulated emission [30].

But the Erbium ions that are excited by the pump laser can also decay back to a lower energy level spontaneously. This amplified spontaneous emission (ASE) reduces the efficiency of the amplifier and generates noise at the receiver. The effect of ASE at the receiver can be described by an additive white Gaussian noise (AWGN) process

2 Fundamentals

$$\underline{E}_{RX}(t) = \underline{E}_{TX}(t) + \underline{n}(t), \qquad (2.23)$$

where $\underline{n}(t)$ is a complex Gaussian random variable with zero mean and variance σ_n^2.

An important measure for optical transmission systems is the optical signal to noise ratio (OSNR), which is defined as the ratio of the signal power P_S to the noise power P_N corrupting the signal [24].

$$\text{OSNR} = \frac{P_S}{P_N} = \frac{P_S}{\rho N_0 B_r} \qquad (2.24)$$

The average noise power is given by the noise power spectral density N_0 within the reference bandwidth B_r. For noise in both polarizations $\rho = 2$, and for noise only in a single-polarization $\rho = 1$. Because of the wide dynamic range of optical signals, the OSNR is usually expressed in a logarithmic decibel scale.

$$\text{OSNR}_{dB} = 10\log_{10}\left(\frac{P_S}{P_N}\right) = 10\log_{10}\left(\frac{P_S}{\rho N_0 B_r}\right) \qquad (2.25)$$

In simulations a normalized signal to noise ratio is used. It is defined as the ratio of the energy per symbol E_S to the noise power spectral density N_0. It is related to the OSNR by the equation

$$\frac{E_S}{N_0} = \frac{P_S}{P_N} \cdot \frac{\rho B_r}{R_S}, \qquad (2.26)$$

where R_S is the symbol rate of the system.

2.2.4 Coherent optical QAM receiver with digital signal processing

2.2.4.1 Polarization diversity coherent optical receiver frontend

Figure 2.9 shows a polarization diversity coherent optical receiver frontend [24; 10]. It consists of a local oscillator lasers, two polarization beam splitters (PBS), two optical 90° hybrids and four differential photodiode pairs with transimpedance amplifiers. It can be considered as an optical down-converter with optical-to-electric conversion. By superimposing the received optical signal with the local oscillator the frequency band of the signal is down-converted into the baseband (homodyne detection) or to an intermediate band with a center frequency at least twice as large as the signal bandwidth (heterodyne detection).

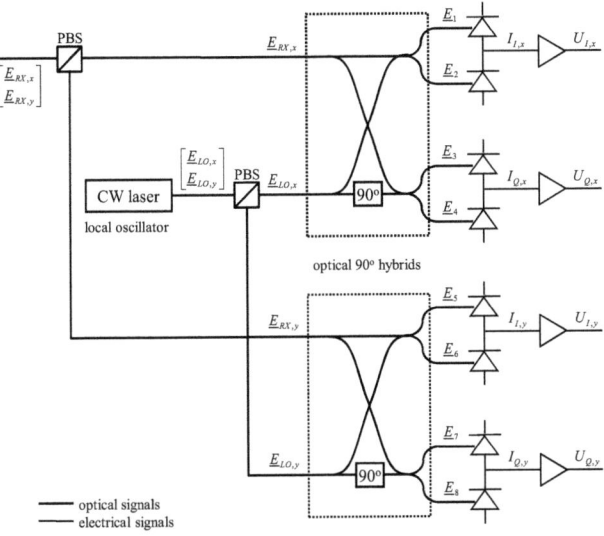

Figure 2.9: Polarization diversity coherent receiver frontend

The receiver input signal is given by the transmitted signal (2.10) multiplied by a Jones matrix **J** (2.17) and corrupted by additive white Gaussian noise.

$$\begin{bmatrix} \underline{E}_{RX,x}(t) \\ \underline{E}_{RX,y}(t) \end{bmatrix} = \mathbf{J}(t) \begin{bmatrix} \underline{E}_{TX,x}(t) \\ \underline{E}_{TX,y}(t) \end{bmatrix} + \begin{bmatrix} \tilde{n}_x(t) \\ \tilde{n}_y(t) \end{bmatrix}$$
$$= \begin{bmatrix} \cos\{\upsilon(t)\}e^{j\delta(t)/2}\underline{E}_{TX,x}(t) - \sin\{\upsilon(t)\}e^{-j\varepsilon(t)/2}\underline{E}_{TX,y}(t) \\ \sin\{\upsilon(t)\}e^{j\varepsilon(t)/2}\underline{E}_{TX,x}(t) + \cos\{\upsilon(t)\}e^{-j\delta(t)/2}\underline{E}_{TX,y}(t) \end{bmatrix} + \begin{bmatrix} \tilde{n}_x(t) \\ \tilde{n}_y(t) \end{bmatrix} \quad (2.27)$$

$\tilde{n}_x(t)$ and $\tilde{n}_y(t)$ are complex Gaussian random variables with zero mean and the variance σ_n^2. Chromatic dispersion and polarization-mode dispersion are not considered yet. The received signal and the local oscillator laser signal

$$\begin{bmatrix} \underline{E}_{LO,x}(t) \\ \underline{E}_{LO,y}(t) \end{bmatrix} = \frac{1}{\sqrt{2}} \begin{bmatrix} \sqrt{2P_{LO}} \\ \sqrt{2P_{LO}} \end{bmatrix} e^{j(\omega_{LO}t + \psi_{LO}(t))} \quad (2.28)$$

with the power P_{LO} and the optical frequency $\omega_{LO}/2\pi$ are split into their two polarization components by two PBS. Then the signals are fed into two optical 90° hybrids. There the signals $\underline{E}_{RX,x}$ and $\underline{E}_{RX,y}$ are superimposed with the four quadrature states in the complex-field space associated with the local oscillator signals $\underline{E}_{LO,x}$ and $\underline{E}_{LO,y}$, respectively. Thus the output signals of the two hybrids are given by

2 Fundamentals

$$\underline{E}_{1,2}(t) = \frac{1}{2}\left(\underline{E}_{RX,x}(t) \pm \frac{1}{\sqrt{2}}\underline{E}_{LO,x}(t)\right),$$

$$\underline{E}_{3,4}(t) = \frac{1}{2}\left(\underline{E}_{RX,x}(t) \pm \frac{j}{\sqrt{2}}\underline{E}_{LO,x}(t)\right),$$
(2.29)

$$\underline{E}_{5,6}(t) = \frac{1}{2}\left(\underline{E}_{RX,y}(t) \pm \frac{1}{\sqrt{2}}\underline{E}_{LO,y}(t)\right),$$

$$\underline{E}_{7,8}(t) = \frac{1}{2}\left(\underline{E}_{RX,y}(t) \pm \frac{j}{\sqrt{2}}\underline{E}_{LO,y}(t)\right).$$
(2.30)

After detection of the outputs of the optical 90° hybrid in differential photoreceivers with the responsivity R and current to voltage conversion in the transimpedance amplifiers with the transfer ratio K, the output voltages of the coherent receiver frontend become

$$U_{I,x}(t) = K\,I_{I,x}(t) = \frac{KR}{2}\left(|\underline{E}_1(t)|^2 - |\underline{E}_2(t)|^2\right) = \frac{KR}{2}\mathrm{Re}\{\underline{E}^+_{RX,x}(t)\underline{E}_{LO,x}(t)\},$$

$$U_{Q,x}(t) = K\,I_{Q,x}(t) = \frac{KR}{2}\left(|\underline{E}_3(t)|^2 - |\underline{E}_4(t)|^2\right) = \frac{KR}{2}\mathrm{Im}\{\underline{E}^+_{RX,x}(t)\underline{E}_{LO,x}(t)\},$$
(2.31)

$$U_{I,y}(t) = K\,I_{I,y}(t) = \frac{KR}{2}\left(|\underline{E}_5(t)|^2 - |\underline{E}_6(t)|^2\right) = \frac{KR}{2}\mathrm{Re}\{\underline{E}^+_{RX,y}(t)\underline{E}_{LO,y}(t)\},$$

$$U_{Q,y}(t) = K\,I_{Q,y}(t) = \frac{KR}{2}\left(|\underline{E}_7(t)|^2 - |\underline{E}_8(t)|^2\right) = \frac{KR}{2}\mathrm{Im}\{\underline{E}^+_{RX,y}(t)\underline{E}_{LO,y}(t)\}.$$
(2.32)

The two output signals of one polarization can be considered as one complex signal. Using (2.12) and (2.17) they are given by

$$\underline{U}_x(t) = U_{I,x}(t) + jU_{Q,x}(t) = \frac{KR}{2}\underline{E}^+_{RX,x}(t)\underline{E}_{LO,x}(t)$$
$$= \frac{KR}{2}\left[\left(\cos\{\upsilon(t)\}e^{j\varepsilon(t)/2}\underline{c}_x(t) - \sin\{\upsilon(t)\}e^{-j\varepsilon(t)/2}\underline{c}_y(t)\right)\sqrt{\frac{P_S P_{LO}}{2}}e^{j((\omega_{LO}-\omega_S)t+\psi_{LO}-\psi_S)} + \underline{n}_x(t)\right],$$
(2.33)

$$\underline{U}_y(t) = U_{I,y}(t) + jU_{Q,y}(t) = \frac{KR}{2}\underline{E}^+_{RX,y}(t)\underline{E}_{LO,y}(t)$$
$$= \frac{KR}{2}\left[\left(\cos\{\upsilon(t)\}e^{-j\varepsilon(t)/2}\underline{c}_y(t) + \sin\{\upsilon(t)\}e^{j\varepsilon(t)/2}\underline{c}_x(t)\right)\sqrt{\frac{P_S P_{LO}}{2}}e^{j((\omega_{LO}-\omega_S)t+\psi_{LO}-\psi_S)} + \underline{n}_y(t)\right],$$
(2.34)

where $\omega_{IF} = \omega_S - \omega_{LO}$ and $\varphi_{IF} = \varphi_S - \varphi_{LO}$ are the frequency and phase differences between the signal and local oscillator lasers. $\underline{n}_x(t)$ and $\underline{n}_y(t)$ are different complex Gaussian random variables, but have the same variance σ_n^2.

It can be seen, that $\underline{I}_x(t)$ and $\underline{I}_y(t)$ contain the transmitted data symbols $\underline{c}_x(t)$ and $\underline{c}_y(t)$. However the two polarization channels are mixed and additionally the received constellation is rotating with the intermediate frequency $\omega_{IF}/2\pi$. Therefore a polarization control and an IF carrier recovery are necessary in order to recover the transmitted data.

2.2.4.2 Analog-to-digital conversion and digital signal processing

To be able to process the received data in the digital domain, the four output voltages of the coherent receiver frontend are sampled by analog-to-digital converters (ADC). The sampling rate of the ADCs should be at least as high as the symbol rate of the system. In practical systems either T_S-spaced sampling or $T_S/2$-spaced sampling is implemented.

For T_S-spaced sampling the discrete output signal of the ADCs is given by

$$\begin{bmatrix}\underline{Z}_{x,k}\\ \underline{Z}_{y,k}\end{bmatrix} \propto \begin{bmatrix}\cos\{\upsilon(kT_S)\}e^{j\delta(kT)/2}\underline{c}_x(kT_S)-\sin\{\upsilon(kT_S)\}e^{-j\delta(kT)/2}\underline{c}_y(kT_S)\\ \sin\{\upsilon(kT_S)\}e^{j\delta(kT)/2}\underline{c}_x(kT_S)+\cos\{\upsilon(kT_S)\}e^{-j\delta(kT)/2}\underline{c}_y(kT_S)\end{bmatrix}e^{j(\omega_{IF}kT+\psi_{IF}(kT))}+\begin{bmatrix}\underline{n}_x(kT_S)\\ \underline{n}_y(kT_S)\end{bmatrix}$$

$$:=\begin{bmatrix}\cos\{\upsilon_k\}e^{j\delta_k/2}\underline{c}_{x,k}-\sin\{\upsilon_k\}e^{-j\delta_k/2}\underline{c}_{y,k}\\ \sin\{\upsilon_k\}e^{j\delta_k/2}\underline{c}_{x,k}+\cos\{\upsilon_k\}e^{-j\delta_k/2}\underline{c}_{y,k}\end{bmatrix}e^{j(\omega_{IF}kT+\psi_{IF,k})}+\begin{bmatrix}\underline{n}_{x,k}\\ \underline{n}_{y,k}\end{bmatrix}.$$

(2.35)

For $T_S/2$-spaced sampling every second sample corresponds to the samples obtained with T_S-spaced sampling. The other samples represent symbol transitions. This so-called oversampling can be advantageous for the compensation of dispersive effects, where the information of one symbol is spread over several samples. Figure 2.10 shows the setup of a coherent digital receiver.

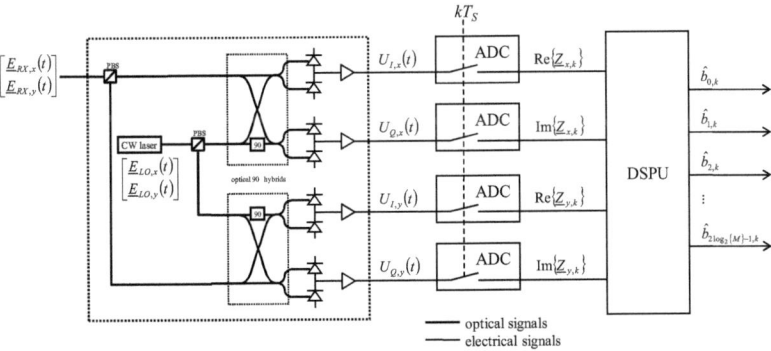

Figure 2.10: Coherent optical receiver with analog-to-digital conversion and digital signal processing

2 Fundamentals

Although it is possible to build a coherent optical receiver with analog signal processing [8; 10], a digital implementation has several considerable advantages.

- Although a manual control of the SOP has successfully been employed in the analog electrical domain [31], an automatic tracking of the SOP has not been demonstrated yet. Therefore an optical polarization control would be required in a commercial system[32]. In contrast in a digital coherent receiver an electronic polarization control with automatic tracking of the SOP is feasible.

- Digital signal processing allows the use of very sophisticated algorithms, e.g. for PMD/CD compensation or mitigation of nonlinear effects. In analog circuits only some simple operations can be realized.

2.2.4.3 Phase noise in a coherent digital receiver

Phase noise originating from both the transmitter and local oscillator laser is a major source of distortion in coherent optical receivers. It is caused by random fluctuations of the instantaneous frequency of the lasers due to their finite linewidth. Typically the linewidth of a laser is specified as the full width at half maximum (FWHM) Δf_{3dB} of its optical power spectrum [33].

The phase fluctuations are described by the so called Wiener-Lévy process [34], which describes a random walk phase modulation. In a coherent digital receiver with discrete variables this random walk is given by

$$\psi_k = \psi_{k-1} + \Delta \psi_k. \quad (2.36)$$

The zero mean Gaussian random variable $\Delta \psi_k$ is referred to as the step-size of the random walk. The fluctuation speed of the process is set by the variance

$$\sigma_\Delta^2 = 2\pi \Delta f_{3dB} T_S \quad (2.37)$$

of $\Delta \psi_k$. Figure 2.11 shows examples for the random walk of ψ_k for different linewidth-times-symbol-duration products $\Delta f_{3dB} T_S$. Figure 2.12 depicts the corresponding Lorentzian carrier power spectra given by

$$P(\Delta f T_S; \Delta f_{3dB} T_S) = \frac{\frac{1}{2} \Delta f_{3dB} T_S}{\pi \left[(\Delta f T_S)^2 + \left(\frac{1}{2} \Delta f_{3dB} T_S \right)^2 \right]}. \quad (2.38)$$

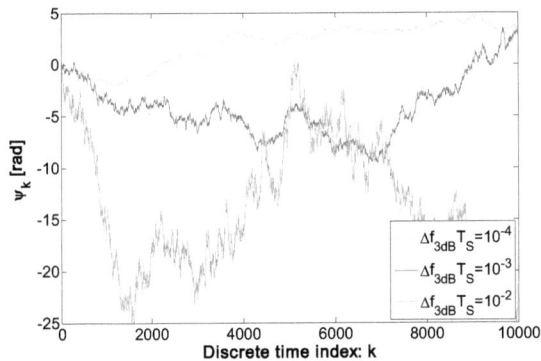

Figure 2.11: Examples of the phase noise process ψ_k for different values of $\Delta f_{3dB}T_S$

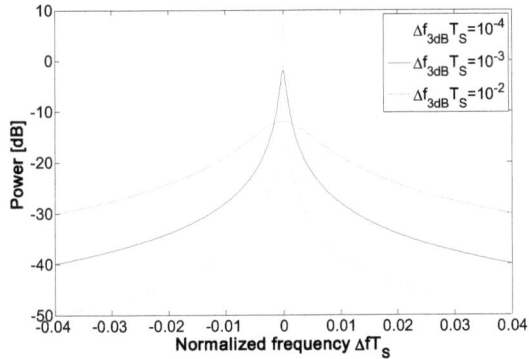

Figure 2.12: Lorentzian carrier power spectra for different values of $\Delta f_{3dB}T_S$

3 Digital signal processing algorithms for coherent optical receivers

This chapter presents the algorithms that are necessary to recover the data in a coherent optical receiver with digital signal processing (Figure 3.1). But as the high data rates in optical communication of 43 Gb/s, 112 Gb/s or even above generate stringent constraints for these algorithms section 3.1 first summarizes these constraints. They are not taken from a book but derived from my personal experience gained during my research. The subsequent algorithms for clock recovery, polarization control, ISI compensation, carrier & data recovery and intermediate frequency control all fulfil these requirements.

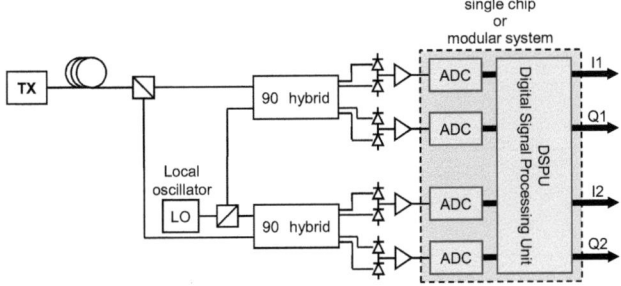

Figure 3.1: Coherent optical receiver structure

3.1 Constraints for algorithms in digital receivers for coherent optical communication

This section summarizes and exemplifies the main real-time constraints for coherent optical receiver algorithms. Three main constraints are identified [4; 12]:

- Feasibility of parallel processing
- Possibility of a hardware-efficient implementation
- Tolerance against feedback delays

The justifications for these constraints are presented in the following sections.

3 Digital signal processing algorithms for coherent optical receivers

3.1.1 Feasibility of parallel processing

Algorithms that support multi-Gb/s operations must allow parallel processing as shown in Figure 3.2 [12]. The DSPU cannot operate directly at the sampling clock frequency of the analog-to-digital converter, which is in general 10 GHz or higher, but requires demultiplexing to process the data in m parallel modules at clock frequencies below 1 GHz. This allows automated generation of the layout, which is indispensable due to the complexity of the system. A comparison between the sampling clock frequency and the divided clock shows that at least $m = 16$ parallel modules are required. Algorithms for real-time applications should therefore theoretically allow parallel processing with an unlimited number of demultiplexed channels. The requirement for this is that (intermediate) results within one module do not depend on results calculated at the same time in other parallel modules.

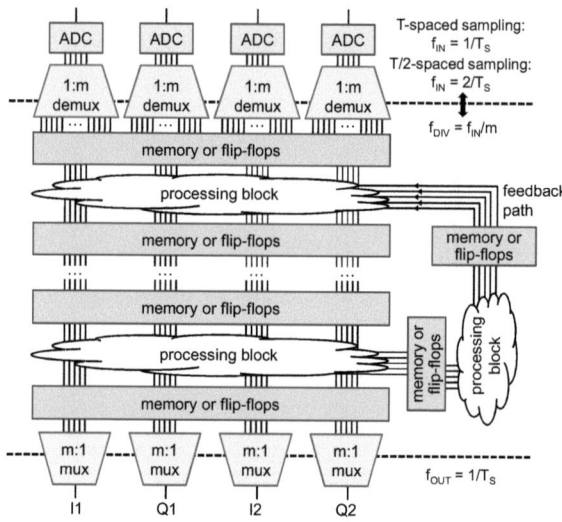

Figure 3.2: Interfacing between ADCs and DSPU and internal structure of the DSPU

A good example to explain the feasibility of parallel processing is the comparison of two filter structures: Finite (FIR) and infinite (IIR) impulse response filters. Figure 3.3 depicts their structures in both serial and parallel systems. It can be seen that it is easily possible to parallelize an FIR filter. Although the output signal

$$y_k = \alpha_0 x_k + \alpha_1 x_{k-1} + \alpha_2 x_{k-2} \quad (3.1)$$

depends on information provided by several parallel modules it does not depend on the result of the same calculation performed in these modules. In contrast, it is a big challenge to realize the parallel structure shown for the IIR filter, because the result

$$y_k = x_k + \beta_1 y_{k-1} + \beta_2 y_{k-2} \quad (3.2)$$

depends on results calculated at the same time in other parallel modules. A very low clock frequency or a low number of parallel channels would be needed to allow all calculations to be executed within one clock cycle. Neither of these requirements is fulfilled in coherent digital receivers for optical transmission system.

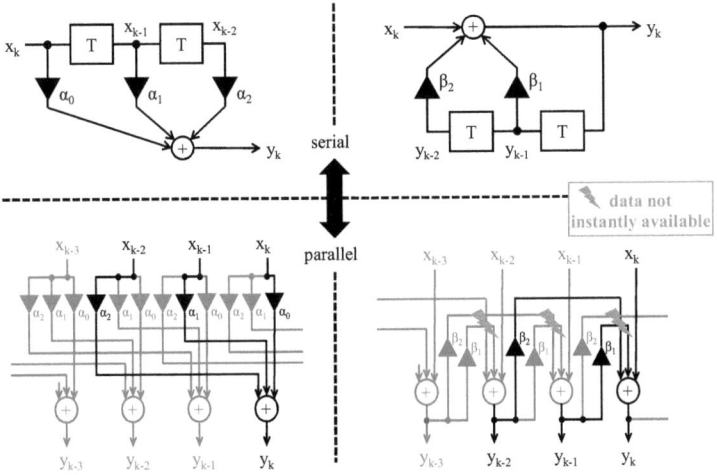

Figure 3.3: Serial and parallel FIR and IIR filter structures

3.1.2 Hardware efficiency

Another important constraint for real-time coherent receiver algorithms is that they must allow for a hardware-efficient implementation. This requirement also originates from the parallel processing in the DSPU. Since most of the required algorithm blocks have to be implemented m times within the DSPU, computationally intensive algorithms require a huge amount of chip area and therefore increase power consumption and cost. The algorithms considered for a chip implementation should therefore not only be evaluated by performance, but also by hardware efficiency.

One way to increase hardware efficiency is to use signal transformations, e.g. FFT/IFFT, log-function or transformation of in-phase and quadrature signal pairs into polar coordinates [35]. Although the transformation itself requires additional hardware effort, this is often beneficial because subsequent calculations are simplified. For example an FFT can reduce a convolution to multiplications, or the log-function or polar coordinates allow replacing multiplications by summations.

The use of look-up tables (LUT) is another effective way to reduce the required hardware. Coordinate transformations and nonlinear functions are the main candidates for a LUT implementation, but it can also be beneficial to replace a multiplication by a LUT, especially if the required precision of the result is low [35].

3 Digital signal processing algorithms for coherent optical receivers

This directly implies the last but most important way to increase efficiency: The optimization of the required precision. In microprocessors and computers it is common to use standard precisions. The MATLAB® programming software for example stores integer variables with 8, 16, 32 or 64 bit precision and floating point variables with 32 or 64 bit precision [IEEE 754-1985]. This is useful to allow the efficient compilation of software. However, in an application-specific hardware design each additional bit increases the required number of gates and hence chip area, power consumption and cost. Therefore all calculations should be optimized to just the precision that is necessary to achieve the required accuracy.

3.1.3 Tolerance against feedback delays

The last important consideration for algorithms for real-time applications is the tolerable feedback delay [4]. In simulation or offline processing feedback delays of 1 symbol are easy to achieve, but this is impossible in a real-time system designed for multi-Gb/s operation. The reasons are the parallel processing and massive pipelining, which is required to cope with the high data rates. Pipelining means that only fractions of the whole algorithm are processed within one clock cycle and the intermediate results are stored in buffers, e.g. memory or flip-flops (FF) as shown in Figure 3.2. Therefore it can take easily 100 symbol durations until a received symbol has an impact on the feedback signal. Algorithms for polarization control and CD/PMD compensation usually employ integral controllers with time constants in the µs-range to update their tap coefficients [12]. In these cases the additional delay due to pipelining, which is in the ns-range, can be neglected. But the feedback delay can have a severe impact on the performance of algorithms that require an instantaneous feedback, e.g. decision-directed carrier recovery, which is often used in offline signal processing for higher-order QAM [36; 37].

Figure 3.4 shows the general structure of a such a decision-directed carrier & data recovery module with an optimum feedback delay $\Delta = 1$ [38]. The phase estimator uses the estimated carrier phase $\hat{\psi}_{IF,k-1}$ to derotate the input symbol $\underline{Y}_k$, and the result is fed into a decision device ($[\ ]_D$ denotes the output of the decision device).

$$\hat{\underline{X}}_k = [\underline{Y}'_k]_D = [\underline{Y}_k \exp\{-j\hat{\psi}_{IF,k-1}\}]_D \tag{3.3}$$

The carrier phase estimate $\hat{\psi}_{IF,k}$ is calculated with

$$\hat{\psi}_{IF,k} = \sum_{i=0}^{N_{CR}} w_i \left(\arg\{\underline{Y}_{k-i}\} - \arg\{\hat{\underline{X}}_{k-i}\}\right), \sum_{i=0}^{N_{CR}} w_i = 1, \ w_i \geq 0, \tag{3.4}$$

where w_i are weighting coefficients (e.g. Wiener filter coefficients, which will be described in detail in section 3.4.2) and N_{CR} is the FIR filter length. If only preceding symbols are used as inputs to the FIR filter, a feedback delay of $\Delta = 1$ becomes theoretically possible [38].

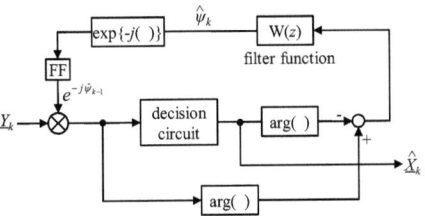

Figure 3.4: Decision-directed carrier recovery with $\Delta = 1$

However, as explained above in a practical implementation it is impossible to achieve $\Delta = 1$. In order to support 100 Gb/s or higher data throughput, practical DSP circuits use massive parallelization and pipelining to realize a synchronous carrier & data recovery as explained in section 3.1.1. Figure 3.5 shows the structure of such an implementation for decision-directed carrier recovery. It can be seen that due to the parallel processing of m consecutive samples the feedback delay between two symbols within one module is equal to the degree of parallelism m. Even if the latest phase estimate from all parallel outputs is fed back into each module as shown in Figure 3.5, the average delay amounts to $(m+1)/2$. Taking also the number of pipeline stages l into account (represented by the flip-flops (FF) in Figure 3.5), which means that it takes l clock cycles until an input sample has an impact on the feedback value, the total average feedback delay is

$$\Delta = (l-1) \cdot m + \frac{m+1}{2}. \tag{3.5}$$

Note that the minimum number of pipeline stages is 1, because this is the minimum number of clock cycles required in a digital feedback system (see Figure 3.4).

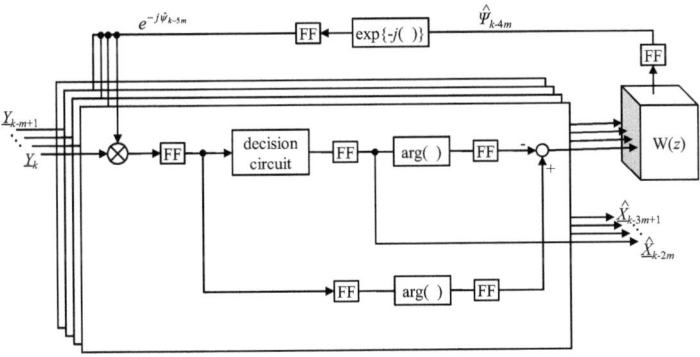

Figure 3.5: Decision-directed carrier recovery in a realistic receiver with parallel and pipelined signal processing.

To determine the effect of Δ on the phase noise tolerance of the receiver, let $\hat{\psi}_{IF,k-\Delta}$ be a perfect estimate of $\psi_{IF,k-\Delta}$, i.e. $\hat{\psi}_{IF,k-\Delta} = \psi_{IF,k-\Delta}$ and $\underline{Y}_k$ is only corrupted by phase noise. With (3.3) $\underline{Y}'_k$ can be written as

$$\underline{Y}'_k = \underline{Y}_k \exp\{j(\psi_{IF,k} - \psi_{IF,k-\Delta})\} = \underline{Y}_k \exp\{j\psi'_{IF,k}\}. \tag{3.6}$$

According to (2.36) ψ'_k is given by

$$\psi'_{IF,k} = \psi_{IF,k} - \psi_{IF,k-\Delta} = \sum_{i=k-\Delta}^{k} \Delta\psi_i . \tag{3.7}$$

$\psi'_{IF,k}$ is a random Gaussian variable with zero mean and variance $\sigma_\psi^2 = \Delta \cdot \sigma_\Delta^2 = 2\pi(\Delta \cdot \Delta f \cdot T_S)$. Therefore the standard deviation of the phase noise increases by the factor $\sqrt{\Delta}$.

In a practical implementation, assuming e.g. parallel processing with $m = 32$ and $l = 5$ pipeline stages, the average feedback delay is $\Delta = 144.5$ samples. This shows that a decision-directed carrier recovery is fairly unfeasible because the phase noise tolerance is reduced by a factor of ~12. Consequently any feedback loop must be avoided in the carrier recovery process, especially for higher-order QAM constellations with their inherently smaller phase noise tolerance. In particular, the carrier cannot be recovered in a decision-directed manner when normal DFB lasers are employed.

A possibility to avoid feedback loops is to apply a feed-forward structure. Several feed-forward carrier recovery schemes will be presented in section 3.4.

3.2 Clock recovery

Clock recovery at the receiver is essential for any data transmissions system, as otherwise a recovery of the transmitted data becomes impossible. Therefore a PLL is required at the receiver to lock the receiver clock phase $\varphi_{CLK,RX}$ to the phase of the transmitter clock $\varphi_{CLK,TX}$. This is implemented using a voltage-controlled oscillator (VCO) as receiver clock source. It is controlled using a clock phase error signal. A common way to generate this clock phase error signal $e_{CLK,k}$ is exploiting the correlation of consecutive samples in an oversampled signal (e.g. $T_S/2$-spaced sampling) [39]. By calculating

$$e_{CLK,k} = |\underline{Z}_{k-1}\underline{Z}_k| - |\underline{Z}_k\underline{Z}_{k+1}| \tag{3.8}$$

the expectation of $e_{CLK,k}$ for a trailing receiver clock phase is $\langle e_{CLK,k} \rangle_{\varphi_{CLK,RX} < \varphi_{CLK,TX}} < 0$, and it is $\langle e_{CLK,k} \rangle_{\varphi_{CLK,RX} > \varphi_{CLK,TX}} > 0$ for a leading receiver clock phase. Therefore the receiver VCO can be locked to the transmitter clock phase by integrating $e_{CLK,k}$ and using it as a control signal for the VCO.

The description for the clock phase error signal generation is only described for single-polarization transmission. An extension to two polarizations is straightforward. However, in a practical implementation an even simpler error signal calculation using only the sign of either the inphase or quadrature channel of one polarization is sufficient.

$$e_{\text{CLK},k} = \text{sgn}[\text{Re}\{\underline{Z}_{k-1}\}]\text{sgn}[\text{Re}\{\underline{Z}_k\}] - \text{sgn}[\text{Re}\{\underline{Z}_k\}]\text{sgn}[\text{Re}\{\underline{Z}_{k+1}\}] \quad (3.9)$$

This significantly reduces the required hardware effort for the error signal calculation, because only operations on bits are required rather than on the whole sample.

3.3 Polarization control & equalization

The purpose of a polarization control circuit is the compensation of cross-talk between the two transmitted polarization channels. The cross-talk between these channels is described by the fiber Jones matrix introduced in section 2.2.3.2. Therefore compensation requires multiplication of the received signal with the estimated inverse of the Jones matrix $\mathbf{M}_k$.

$$\begin{bmatrix} \underline{Y}_{k,x} \\ \underline{Y}_{k,y} \end{bmatrix} = \mathbf{M}_k \begin{bmatrix} \underline{Z}_{k,x} \\ \underline{Z}_{k,y} \end{bmatrix} \quad (3.10)$$

In order to obtain $\mathbf{M}_k$ a non-data-aided (NDA) approach can be employed, where the input and output data of the polarization controller is used to estimate the Jones matrix. Another option is to use a decision-directed (DD) approach where the data before and behind the carrier & data recovery is used for the estimation process.

But not only polarization cross-talk corrupts the signal. Dispersive effects such as chromatic dispersion (CD) (section 2.2.3.3) or polarization mode dispersion (section 2.2.3.4) add additional intersymbol interference, i.e. cross-talk between consecutive symbols. In order to compensate for these effects a novel algorithm is presented in this book, which is based on the principles used for the decision-directed polarization control.

3.3.1 Non-data-aided polarization control

To demultiplex the two polarization channels in a non-data-aided approach the constant modulus algorithm (CMA) is used. The algorithm exploits the fact that polarization cross-talk causes the amplitude of the signal to fluctuate. Thus by minimizing these fluctuations and forcing the signal on the unity circle a perfect separation of two polarization channels can be achieved. The required error signal is calculated as

$$\mathbf{T}_k = \begin{bmatrix} \left(1 - |\underline{Y}_{k,x}|^2\right)\underline{Y}_{k,x} \\ \left(1 - |\underline{Y}_{k,y}|^2\right)\underline{Y}_{k,y} \end{bmatrix} \cdot \begin{bmatrix} \underline{Z}_{k,x} \\ \underline{Z}_{k,y} \end{bmatrix}^+, \quad (3.11)$$

and the polarization control matrix is incrementally updated by

$$\mathbf{M}_k = \mathbf{M}_{k-1} + g\mathbf{T}_{k-\Delta}. \tag{3.12}$$

Figure 3.6 depicts the CMA graphically. A detailed derivation of the CMA can be found in [40], its adaptation for polarization control in a coherent optical receiver is explained in [41].

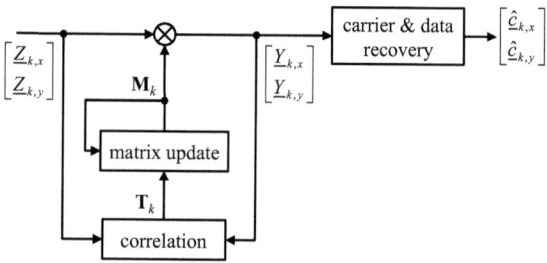

Figure 3.6: Non-data-aided polarization control algorithm

A disadvantage of this algorithm is that in general $\langle \mathbf{M}_k \rangle \neq \mathbf{J}_k^{-1}$, because the CMA only minimizes the cross-talk between the two polarization channels but does not recover the phase offset between them. To be able to achieve also a phase alignment between the two polarization modes an additional control circuit needs to be added or as an alternative a decision-directed polarization control algorithm can be applied.

3.3.2 Decision-directed polarization control

To estimate the Jones matrix $\mathbf{J}$ in a decision-directed approach the recovered symbol $[\hat{c}_{k,x} \ \hat{c}_{k,y}]^T$ must be correlated with the output signal of the polarization control block [12]. Using equation (2.35) and (3.10) the output signal of the polarization controller is

$$\begin{bmatrix} \underline{Y}_{k,x} \\ \underline{Y}_{k,y} \end{bmatrix} \propto \mathbf{M}_k \mathbf{J}_k \begin{bmatrix} \underline{c}_{k,x} \\ \underline{c}_{k,y} \end{bmatrix} e^{j\psi_{IF,k}} + \begin{bmatrix} \underline{n}_{k,x} \\ \underline{n}_{k,y} \end{bmatrix}. \tag{3.13}$$

The correlation matrix $\mathbf{Q}_k$ is therefore given by

$$\mathbf{Q}_k = \begin{bmatrix} \underline{Y}_{k,x} \\ \underline{Y}_{k,y} \end{bmatrix} e^{-j\hat{\varphi}_k} \begin{bmatrix} \hat{c}_{k,x} \\ \hat{c}_{k,y} \end{bmatrix}^+, \tag{3.14}$$

where $\hat{\varphi}_k$ is the estimated carrier phase. The expectation $\langle \mathbf{Q}_k \rangle$ of the matrix $\mathbf{Q}_k$ is a perfect estimate of the matrix product $\mathbf{M}_k \mathbf{J}_k$.

$$\langle \mathbf{Q}_k \rangle = \mathbf{M}_k \mathbf{J}_k \tag{3.15}$$

Therefore by calculating

$$\mathbf{M}_k := \langle \mathbf{Q}_{k-\Delta} \rangle^{-1} \mathbf{M}_{k-\Delta} = \mathbf{J}_{k-\Delta}^{-1} \mathbf{M}_{k-\Delta}^{-1} \mathbf{M}_{k-\Delta} = \mathbf{J}_{k-\Delta}^{-1}, \tag{3.16}$$

where Δ is the processing delay of the system introduced in section 3.1.3, the polarization can be controlled electronically and penalty-free if $\Delta \mathbf{J}_k = \mathbf{J}_k - \mathbf{J}_{k-\Delta} \to 0$. Figure 3.7 visualizes the structure of the algorithm.

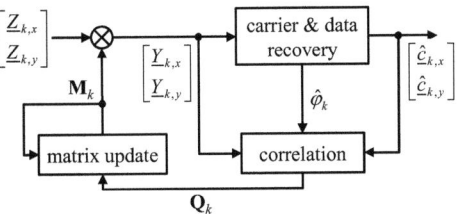

Figure 3.7: Decision-directed polarization control algorithm

The implementation of equation (3.16) poses several challenges. First to calculate the inverse of a matrix in a digital circuit is very complex. But the calculation can be simplified by using the Taylor series [42]

$$\mathbf{Q}^{-1} = \sum_{i=0}^{\infty} (-1)^i (\mathbf{Q}-\mathbf{1})^i. \qquad (3.17)$$

If $\mathbf{Q} \to \mathbf{1}$, which is the goal of the polarization control, the Taylor series is dominated by its first order element and $\mathbf{M}_k$ can be calculated by

$$\mathbf{M}_k := (1 + 1 - \langle \mathbf{Q}_{k-\Delta} \rangle) \mathbf{M}_{k-\Delta} = \mathbf{M}_{k-\Delta} + (1 - \langle \mathbf{Q}_{k-\Delta} \rangle) \mathbf{M}_{k-\Delta}. \qquad (3.18)$$

The second problem is that in general $\langle \mathbf{Q}_k \rangle \neq \mathbf{Q}_k$ applies. Therefore the update algorithm for $\mathbf{M}_k$ must be extended:

$$\mathbf{M}_k := \mathbf{M}_{k-(W-1)-\Delta} + \frac{g}{W} \left(\sum_{w=0}^{W-1} (1 - \mathbf{Q}_{k-w-\Delta}) \right) \mathbf{M}_{k-(W-1)-\Delta}. \qquad (3.19)$$

If $g = 1$ then the expectation $\langle \mathbf{Q}_k \rangle$ is calculated by averaging over W correlations. Therefore $\mathbf{M}_k$ can be updated every W clock cycles. If $W = 1$ then $\mathbf{M}_k$ is updated incrementally every clock cycle using directly the correlation result $\mathbf{Q}_{k-\Delta}$ multiplied with the control gain g. The control time constant c_t is given by

$$c_t = \frac{W}{g} T_S. \qquad (3.20)$$

This means that within the time c a control error decays to the $1/e$-fold error. The ratio W/g should be in the range of $[10^2, 10^4]$, depending on whether an accurate or a fast polarization control should be realized.

3 Digital signal processing algorithms for coherent optical receivers

In principle, the polarization control could also recover the carrier, but the time constant c is much too large to be able to track the phase noise caused by standard DFB lasers. Therefore an additional feed-forward carrier recovery must be applied.

3.3.3 Decision-directed ISI compensation

The decision-directed polarization control algorithm presented in section 3.3.2 can be extended to additionally allow for intersymbol interference (ISI) compensation caused e.g. by PMD. Therefore equation (3.10) is changed to the following FIR filter structure:

$$\begin{bmatrix} Y_{k,x} \\ Y_{k,y} \end{bmatrix} = \sum_{i=-N_{PMDC}}^{N_{PMDC}} \mathbf{M}_{i,k} \begin{bmatrix} Z_{k-i,x} \\ Z_{k-i,y} \end{bmatrix}. \qquad (3.21)$$

N_{PMDC} is the FIR filter half width, i.e. $2N_{PMDC}+1$ input vectors are weighted by the compensation matrices $\mathbf{M}_i$ and summed up.

To update the filter matrices the correlations between the filter output vectors of the N_{PMDC} preceding and subsequent symbols are correlated with the current recovered symbol similar to equation (3.14).

$$\mathbf{Q}_{n,k} = \left(\left(\frac{1}{2} + \chi_n \right) \begin{bmatrix} Y_{k-n,x} \\ Y_{k-n,y} \end{bmatrix} - \chi_n \begin{bmatrix} Y_{k,x} \\ Y_{k,y} \end{bmatrix} \right) e^{-j\hat{\varphi}_{k-n}} \begin{bmatrix} \hat{c}_{k,x} \\ \hat{c}_{k,y} \end{bmatrix}^+ \quad |n| = 1, \ldots, N_{PMDC} \qquad (3.22)$$

The parameter χ_n compensates for inherent correlations between neighboring symbols and depends on the applied oversampling. For $T_S/2$-spaced sampling $\chi_{\pm 1} = \frac{1}{2}$, all other χ_n, $|n| \neq 1$, are zero.

The expectations $\langle \mathbf{Q}_{n,k} \rangle$, $n \neq 0$ of the correlation results are perfect estimates of the ISI between the symbols with the indices k and $k-n$. More precise the expectation $\langle \mathbf{Q}_{n,k} \rangle$ is a perfect estimate of the matrix that describes how the symbol $[X_{k,x} \ X_{k,y}]^T$ is coupled into the symbol $[X_{k-n,x} \ X_{k-n,y}]^T$. Polarization crosstalk within the center symbol can be compensated by using equation (3.18). Therefore if $[Y_{k,x} \ Y_{k,y}]^T$ were decoded a second time ISI and polarization crosstalk could be significantly reduced by calculating

$$\begin{bmatrix} \tilde{Y}_{k,x} \\ \tilde{Y}_{k,y} \end{bmatrix} = \begin{bmatrix} Y_{k,x} \\ Y_{k,y} \end{bmatrix} + \left(1 - \langle \mathbf{Q}_{0,k} \rangle \right) \begin{bmatrix} Y_{k,x} \\ Y_{k,y} \end{bmatrix} - \sum_{n=1}^{N_{PMDC}} \left(\langle \mathbf{Q}_{-n,k} \rangle \begin{bmatrix} Y_{k-n,x} \\ Y_{k-n,y} \end{bmatrix} + \langle \mathbf{Q}_{n,k} \rangle \begin{bmatrix} Y_{k+n,x} \\ Y_{k+n,y} \end{bmatrix} \right). \qquad (3.23)$$

By setting equation (3.21) into equation (3.23) and neglecting all input samples $[Z_{k-i,x} \ Z_{k-i,y}]^T$ with $|i| > N_{PMDC}$ one obtains

$$\begin{bmatrix} \tilde{Y}_{k,x} \\ \tilde{Y}_{k,y} \end{bmatrix} = \sum_{i=-N_{PMDC}}^{N_{PMDC}} \left(\mathbf{M}_{i,k} + \tilde{\mathbf{M}}_{i,k} \right) \begin{bmatrix} Z_{k-i,x} \\ Z_{k-i,y} \end{bmatrix}, \qquad (3.24)$$

where the $\widetilde{\mathbf{M}}_{i,k}$ are given by

$$\widetilde{\mathbf{M}}_{i,k} = \begin{cases} \left(1 - \langle \mathbf{Q}_{0,k} \rangle\right)\mathbf{M}_{i,k} - \sum_{n=1}^{N_{\text{PMDC}}}\langle \mathbf{Q}_{n,k}\rangle\mathbf{M}_{n+i,k} - \sum_{n=-N_{\text{PMDC}}-i}^{-1}\langle \mathbf{Q}_{n,k}\rangle\mathbf{M}_{n+i,k} & i < 0 \\ \left(1 - \langle \mathbf{Q}_{0,k} \rangle\right)\mathbf{M}_{0,k} - \sum_{n=1}^{N_{\text{PMDC}}}\langle \mathbf{Q}_{n,k}\rangle\mathbf{M}_{n,k} - \sum_{n=-N_{\text{PMDC}}}^{-1}\langle \mathbf{Q}_{n,k}\rangle\mathbf{M}_{n,k} & i = 0 \\ \left(1 - \langle \mathbf{Q}_{0,k} \rangle\right)\mathbf{M}_{i,k} - \sum_{n=1}^{N_{\text{PMDC}}-i}\langle \mathbf{Q}_{n,k}\rangle\mathbf{M}_{n+i,k} - \sum_{n=-N_{\text{PMDC}}}^{-1}\langle \mathbf{Q}_{n,k}\rangle\mathbf{M}_{n+i,k} & i > 0 \end{cases} \quad (3.25)$$

Taking into account that in general $\langle \mathbf{Q}_{n,k}\rangle \neq \mathbf{Q}_{n,k}$ holds in a practical system, the compensation matrices can be updated by

$$\mathbf{M}_{i,k} = \mathbf{M}_{i,k-\Delta} + \frac{g}{W}\widetilde{\mathbf{M}}'_{i,k-\Delta} \quad |i| = 1, \ldots, N_{\text{PMDC}} \quad (3.26)$$

with

$$\widetilde{\mathbf{M}}'_{i,k} = \begin{cases} \left(\sum_{w=0}^{W-1}\left(1 - \mathbf{Q}_{0,k-w}\right)\right)\mathbf{M}_{i,k} - \sum_{n=1}^{N_{\text{PMDC}}}\left(\sum_{w=0}^{W-1}\mathbf{Q}_{-n,k-w}\right)\mathbf{M}_{n+i,k} - \sum_{n=-N_{\text{PMDC}}-i}^{-1}\left(\sum_{w=0}^{W-1}\mathbf{Q}_{n,k-w}\right)\mathbf{M}_{n+i,k} & i < 0 \\ \left(\sum_{w=0}^{W-1}\left(1 - \mathbf{Q}_{0,k-w}\right)\right)\mathbf{M}_{0,k} - \sum_{n=1}^{N_{\text{PMDC}}}\left(\sum_{w=0}^{W-1}\mathbf{Q}_{-n,k-w}\right)\mathbf{M}_{n,k} - \sum_{n=-N_{\text{PMDC}}}^{-1}\left(\sum_{w=0}^{W-1}\mathbf{Q}_{n,k-w}\right)\mathbf{M}_{n,k} & i = 0 \\ \left(\sum_{w=0}^{W-1}\left(1 - \mathbf{Q}_{0,k-w}\right)\right)\mathbf{M}_{i,k} - \sum_{n=1}^{N_{\text{PMDC}}-i}\left(\sum_{w=0}^{W-1}\mathbf{Q}_{-n,k-w}\right)\mathbf{M}_{n+i,k} - \sum_{n=-N_{\text{PMDC}}}^{-1}\left(\sum_{w=0}^{W-1}\mathbf{Q}_{n,k-w}\right)\mathbf{M}_{n+i,k} & i > 0 \end{cases}$$

(3.27)

The derived update algorithm allows to efficiently compensate for ISI caused by CD or PMD. But updating the compensation matrices is computationally very expensive because $3N_{\text{PMDC}}(N_{\text{PMDC}}+1)+1$ complex matrix multiplications are required. In order to reduce the complexity of the updating process it is possible to consider only the dominant summand in the calculation for each $\widetilde{\mathbf{M}}_{i,k}$. To identify this summand the energy distribution of the spread symbol can be investigated. Most of the energy is contained in the center symbol and it decays towards both directions. Thus each compensation matrix $\widetilde{\mathbf{M}}_{i,k}$ is dominated by the summand $\langle \mathbf{Q}_{-i,k}\rangle\mathbf{M}_{0,k}$. If the other summands are neglected equation (3.25) reduces to

$$\widetilde{\mathbf{M}}_{i,k} = \begin{cases} -\langle \mathbf{Q}_{-i,k}\rangle\mathbf{M}_{0,k} & i \neq 0 \\ \left(1 - \langle \mathbf{Q}_{0,k}\rangle\right)\mathbf{M}_{0,k} & i = 0 \end{cases}. \quad (3.28)$$

Still all $\langle \mathbf{Q}_{-i,k}\rangle$, $i \neq 0$ are forced towards zero. Hence no performance degradation has to be expected compared to the updating process given in equation (3.25). Another way to derive equation (3.28) is also described in [43].

3.4 Feed-forward carrier recovery

After polarization control and intermediate frequency compensation, which will be described in section 3.6, the input signal into the carrier recovery block is given by

$$\begin{bmatrix} \underline{Y}_{x,k} \\ \underline{Y}_{y,k} \end{bmatrix} = \begin{bmatrix} |\underline{Y}_{x,k}| e^{j\varphi_{x,k}} \\ |\underline{Y}_{y,k}| e^{j\varphi_{y,k}} \end{bmatrix} \propto \begin{bmatrix} \underline{c}_{x,k} \\ \underline{c}_{y,k} \end{bmatrix} e^{j(\psi_{IF,k})} + \begin{bmatrix} \underline{n}_{x,k} \\ \underline{n}_{y,k} \end{bmatrix} \quad (3.29)$$

with

$$\begin{bmatrix} \varphi_{x,k} \\ \varphi_{y,k} \end{bmatrix} = \begin{bmatrix} \arg\{\underline{c}_{x,k}\} + \psi_{IF,k} + n'_{x,k} \\ \arg\{\underline{c}_{y,k}\} + \psi_{IF,k} + n'_{y,k} \end{bmatrix}. \quad (3.30)$$

The input signal $\underline{Y}_k$ is sampled at the symbol rate, and perfect intermediate frequency compensation and equalization are assumed.

As the polarizations are perfectly separated and the carrier recovery is identical for both polarizations in the following the algorithms are described for a single-polarization system and the indices denoting the polarization channel are omitted.

3.4.1 Viterbi & Viterbi algorithm

The Viterbi & Viterbi algorithm allows feed-forward carrier recovery for constellations with equidistant symbol phases [44]. If the phase offset between two adjacent symbols is $2\pi/p$ the modulation can be eliminated by raising the input signal $\underline{Y}_k$ to the pth power. In the case of QPSK $p = 4$.

$$\underline{U}_k = |\underline{Y}_k|^u e^{jp\varphi_k} \quad u \in \{0, 2, 4\} \quad (3.31)$$

Due to $\arg\{\underline{c}_k^p\} = \pi$ the result is independent of modulation-induced phase changes [9]. The factor u allows to choose whether $\underline{U}_k$ should depend on the amplitude $|\underline{Y}_k|$ or not. For $u = p$ equation (3.31) equates to $\underline{U}_k = \underline{Y}_k^p$, for $u = 0$ it can be rewritten as $\underline{U}_k = e^{jp\varphi_k}$.

The resulting frequency multiplied carrier components are then filtered to remove distortions caused by noise.

$$\underline{V}_k = \frac{1}{2N_{CR} + 1} \sum_{n=-N_{CR}}^{N_{CR}} \underline{U}_{k-n} \quad (3.32)^1$$

---

[1] For a joined carrier recovery for both polarizations equation (3.32) must be extended to

$$\underline{V}_k = \frac{1}{2(2N_{CR} + 1)} \sum_{n=-N_{CR}}^{N_{CR}} (\underline{U}_{k-n,x} + \underline{U}_{k-n,y})$$

N_{CR} is referred to as filter half width, i.e. $2N_{CR}+1$ values are summed up to calculate $\underline{V}_k$. $N_{CR}=0$ is equivalent to asynchronous demodulation of the signal.

The filter also alters the phase angle: $\underline{V}_k \propto -e^{jp\hat{\psi}_{IF,k}}$. In the noise-free case $p\hat{\psi}_{IF,k} = p\psi_{IF,k}$. To finally recover the carrier phase the phase of $\underline{V}_k$ must be divided by a factor of p.

$$\hat{\psi}_{IF,k} = \frac{1}{p}\arg(\underline{V}_k) \qquad (3.33)$$

Figure 3.8 shows the schematic of the Viterbi & Viterbi carrier recovery algorithm.

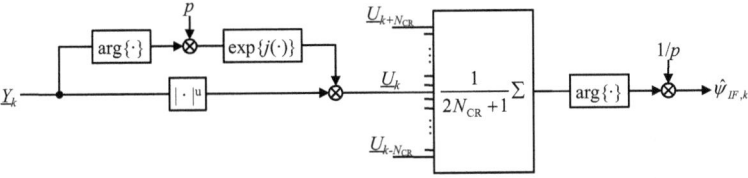

Figure 3.8: Viterbi & Viterbi feed-forward carrier recovery

3.4.2 Weighted Viterbi & Viterbi algorithm

Looking again at the statistical properties of ψ_{IF} given by equation (2.36) it can be seen that most information about $\psi_{IF,k}$ is contained in the symbol k itself and that it is symmetrically decaying for symbols with increasing distance to k. As in the filter function (3.32) of the original Viterbi & Viterbi algorithm all symbols are weighted with the same weight $1/(2N_{CR}+1)$, it is only optimal for $\sigma_\Delta^2 = 0$. It can be optimized by using a Wiener filter with variable weights that reflect the information content about $\psi_{IF,k}$ in the weighted symbol [34].

$$\underline{V}_k = \sum_{n=-N_{CR}}^{N_{CR}} v_n \underline{U}_{k-n} \qquad (3.34)^2$$

The optimal values for the Wiener coefficients v_n depend on the ratio of laser-induced phase noise to the angular portion of AWGN $\sigma_\Delta^2/\sigma_{n'}^2$ and are given by [38]

$$v_n = \frac{\kappa}{1-\kappa^2}\frac{\sigma_\Delta^2}{\sigma_{n'}^2}\kappa^{|n|} \qquad (3.35)$$

---

[2] For a joined carrier recovery for both polarizations equation (3.34) must be extended to
$$\underline{V}_k = \sum_{n=-N_{CR}}^{N_{CR}} v_n \left(\underline{U}_{k-n,x} + \underline{U}_{k-n,y}\right)$$

with

$$\kappa = \left(1 + \frac{1}{2}\frac{\sigma_\Delta^2}{\sigma_{n'}^2}\right) - \sqrt{\left(1 + \frac{1}{2}\frac{\sigma_\Delta^2}{\sigma_{n'}^2}\right)^2 - 1}. \quad (3.36)$$

The sum of the Wiener coefficients is $\sum_{n=-\infty}^{\infty} v_n = 1$. Figure 3.9 shows the modified Viterbi & Viterbi algorithm containing the additional weighting factors.

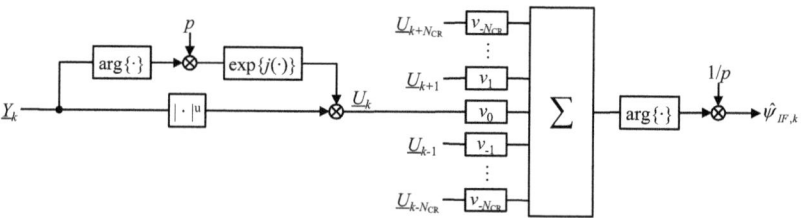

Figure 3.9: Weighted Viterbi & Viterbi feed-forward carrier recovery

The value $\sigma_\Delta^2/\sigma_{n'}^2$ can be estimated in the receiver by comparing the noise variances in the tangential and radial directions.

$$\frac{\sigma_\Delta^2}{\sigma_{n'}^2} = \frac{\left\langle \left(\varphi_k - \hat{\psi}_{IF,k}\right)^2 \right\rangle}{\left\langle \left(\frac{|Y_k|}{\langle |Y_k| \rangle} - 1\right)^2 \right\rangle} - 1 \quad (3.37)$$

As $\sigma_\Delta^2/\sigma_{n'}^2$ is in general not time-variant its estimation can be done by software with a slow update rate. Another option is it to omit the estimation of $\sigma_\Delta^2/\sigma_{n'}^2$ and set it to a constant value.

A further increase in hardware efficiency can be achieved if the Wiener coefficients v_n are rounded to multiples of 2^{-a}, $a \in \{0,1,2,...\}$. In a digital circuit this allows to realize the calculation of the products $v_n \underline{U}_{k-n}$ by using simple shift-and-add operations instead of multipliers.

3.4.3 Barycenter algorithm

The Viterbi & Viterbi algorithm recovers the carrier phase using operations on complex numbers. However the amplitudes of these numbers contain only limited information about the carrier phase, if any. Therefore it is straightforward to estimate the carrier angle ψ_{IF} in the angular domain using only $\varphi_k = \arg\{\underline{Y}_k\}$. This algorithm is referred to as barycenter algorithm [45] or maximum likelihood phase approximation (MLPA) [46].

In a first step modulation-induced phase changes are removed. If the constellation is rotationally symmetric by the angle $2\pi/p$, a modulation-free symbol phase is obtained by calculating

$$\vartheta_k = \varphi_k \mod \frac{2\pi}{p}. \tag{3.38}$$

This operation is similar to Viterbi & Viterbi carrier recovery with $u = 0$ and $N_{CR} = 0$, because it equates to $\vartheta_k = (1/p)\arg\{Y_k^p\}$ with $\vartheta_k \in [0, 2\pi/p)$.

To remove distortions caused by the angular portion of AWGN the ϑ_k need to be filtered. But in contrast to filtering in the complex plane simple averaging over several consecutive phase angles is not possible due to the limited co-domain of ϑ_k. To overcome this problem the statistical properties of ψ_{IF} can be exploited. Therefore the filter is constructed in a cellular approach, in which each cell performs the following function [47]:

$$\phi_k = \begin{cases} \left(\dfrac{\theta_k + \zeta_k}{2} - \dfrac{\pi}{p}\right) \mod \dfrac{2\pi}{p} & \theta_k - \zeta_k \geq \dfrac{\pi}{p} \\ \dfrac{\theta_k + \zeta_k}{2} \mod \dfrac{2\pi}{p} & -\dfrac{\pi}{p} \leq \theta_k - \zeta_k < \dfrac{\pi}{p} \\ \left(\dfrac{\theta_k + \zeta_k}{2} + \dfrac{\pi}{p}\right) \mod \dfrac{2\pi}{p} & \theta_k - \zeta_k < \dfrac{\pi}{p} \end{cases} \tag{3.39}$$

θ_k and ζ_k are the input angles of the filter cell. Similar to the differential decoding process described in section 3.4.1, always the shortest possible physical path between θ_k and ζ_k is chosen for calculation of the mean angle ϕ_k. Larger filter structures can be designed by combining several filter cells in a tree structure. Depending on the arrangement of the filter cells several filter structures are possible for the same filter half width N_{CR}. Table 3.1 shows the structures that are examined in this book.

3 Digital signal processing algorithms for coherent optical receivers

Table 3.1: Barycenter carrier recovery filter structures

N_{CR}	Filter structure[3]
1	
2	
3	
4	
5	

---

[3] For a joined carrier recovery for both polarizations the outputs from two filters $\hat{\psi}_{IF,k,x}$ and $\hat{\psi}_{IF,k,y}$ can be combined in an additional filter cell to determine the joined carrier phase $\hat{\psi}_{IF,k}$.

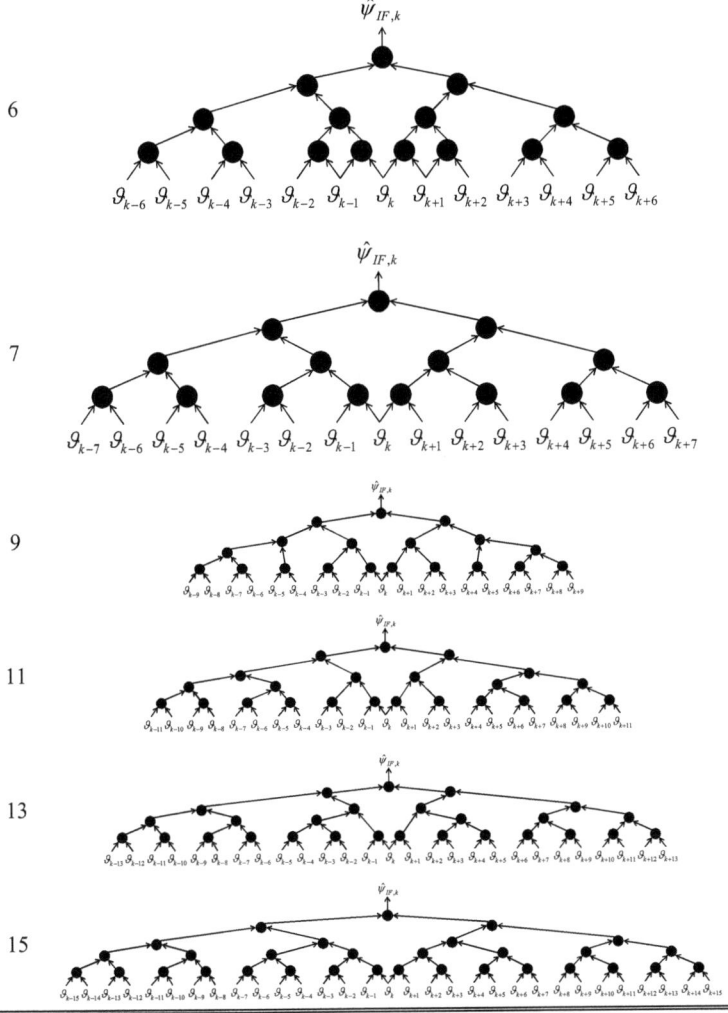

A closer look at the structures depicted in Table 3.1 unveils that the barycenter algorithm also incorporates a weighting of the input coefficients. For example in the filter with $N_{CR} = 1$ the center input is fed twice into the filter and is therefore also weighted twice compared to the other inputs. For $N_{CR} = 13$ the center input is weighted four times and its neighbour inputs twice as strong as the other inputs.

Using the final filter output, which represents the estimated carrier phase $\hat{\psi}_{IF,k}$, the transmitted data can be recovered using equation (3.46).

3 Digital signal processing algorithms for coherent optical receivers

Problematic for the performance of the filter is the case if for one of the filter cells $|\theta_k - \zeta_k| \approx \pi/p$. In this case two results have roughly the same probability. Hence the output of the filter cell is highly unreliable and might severely falsify the overall filter output. This can be avoided by adding reliability information to each intermediate result. Then a subsequent filter cell can decide on the basis of this reliability information, if it is advantageous to discard an unreliable input and just feed-through the other one, or in case of two reliable inputs to perform the averaging function according to equation (3.39). This improved barycenter algorithm is referred to as selective maximum likelihood phase approximation (SMLPA) [46].

3.4.4 Feed-forward carrier recovery for arbitrary QAM constellations

The described feed-forward carrier recovery concepts until now are only able to efficiently recover the carrier phase for QAM constellations with equidistant-phases. But their performance for square or other possible QAM constellations is very poor [48]. They either require averaging over a large number of symbols ($N_{CR} > 100$) or to use only dedicated symbols for carrier recovery that fulfill the equidistant-phase constraint [49]. Both significantly reduces the phase noise tolerance of the algorithms.

3.4.4.1 Square QAM constellations

In this book a novel feed-forward carrier recovery concept is presented that is able to recover the carrier phase from arbitrary QAM constellations. Figure 3.10 shows a block diagram of the proposed carrier recovery for square QAM constellations. As these constellations are the most important for practical systems the derivation of the concept is first presented for square QAM constellations. Afterwards the concept will be generalized to arbitrary constellations.

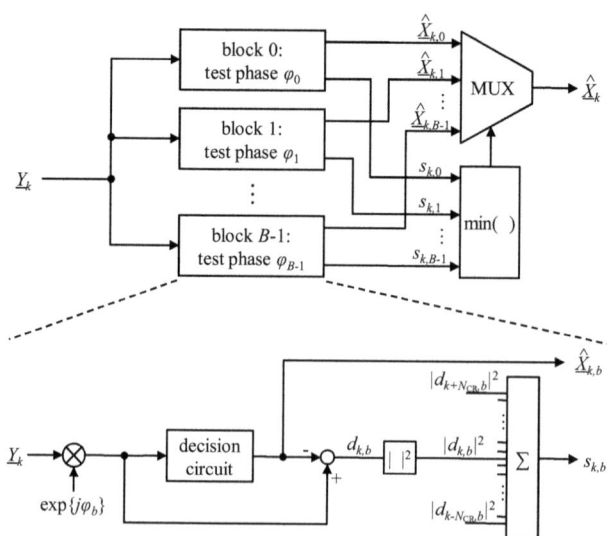

Figure 3.10: Feed-forward carrier recovery for square QAM constellations

The input signal Z_k of the coherent receiver is sampled at the symbol rate, and perfect clock recovery and equalization are assumed. To recover the carrier phase in a pure feed-forward approach the received signal Z_k is rotated by B test carrier phase angles φ_b with

$$\varphi_b = \frac{b}{B} \cdot \frac{\pi}{2}, \ b \in \{0,1,...,B-1\}. \tag{3.40}$$

Then all rotated symbols are fed into a decision circuit, which will be described in section 3.5.2, and the squared distance $|d_{k,b}|^2$ to the closest constellation point is calculated in the complex plane:

$$\begin{aligned}|d_{k,b}|^2 &= |Z_k \exp\{j\varphi_b\} - \lfloor Z_k \exp\{j\varphi_b\}\rfloor_D|^2 \\ &= |Z_k \exp\{j\varphi_b\} - \hat{X}_{k,b}|^2\end{aligned} \tag{3.41}$$

In order to remove noise distortions, the distances of $2N_{CR}+1$ consecutive test symbols rotated by the same carrier phase angle φ_b are summed up:

$$s_{k,b} = \sum_{n=-N_{CR}}^{N_{CR}} |d_{k-n,b}|^2 \tag{3.42}^4$$

The optimum value of N_{CR} depends on the laser linewidth-times-symbol-duration product and will be evaluated by simulation in section 4.2.2.
After filtering the optimum phase angle is determined by searching the minimum sum of distance values $s_{k,b_{min,k}}$. As the decoding was already executed in (3.41), the recovered output symbol $\hat{X}_k$ can be selected by a switch controlled by the index $b_{min,k}$ of the minimum distance sum:

$$\hat{X}_k := \hat{X}_{k,b_{min,k}} \tag{3.43}$$

3.4.4.2 Arbitrary QAM constellations

The proposed feed-forward carrier recovery concept can also be applied to arbitrary QAM constellations. If the constellation diagram is rotationally symmetric by the angle γ, then φ_b must be selected as

$$\varphi_b = \frac{b}{B} \cdot \gamma, \ b \in \{0,1,...,B-1\}. \tag{3.44}$$

---

[4] For a joined carrier recovery for both polarizations equation (3.42) must be extended to

$$s_{k,b} = \sum_{n=-N_{CR}}^{N_{CR}} \left(|d_{k-n,b,x}|^2 + |d_{k-n,b,y}|^2\right)$$

3 Digital signal processing algorithms for coherent optical receivers

For square QAM constellations $\gamma = \pi/2$ as used in equation (3.40). Without rotational symmetry $\gamma = 2\pi$ must be used.

Due to the k-fold ambiguity of the recovered phase with $k = 2\pi/\gamma$, $\lceil \log_2\{k\} \rceil$ bits should be differentially encoded/decoded, where $\lceil x \rceil$ is the smallest integer larger than or equal to x.

3.4.4.3 Hardware-efficient implementation

Due to the extremely high degree of parallelization (almost each functional block is in total required $B \cdot m$ times, where m is the number of parallel modules due to demultiplexing introduced in section 3.1.1), a highly efficient implementation of the algorithm is required to make it feasible for implementation in a real-time receiver. Therefore this subsection lists some recommendations for a hardware-efficient implementation.

3.4.4.3.1 Efficient calculation of vector rotations

The rotation of a symbol in the complex plane normally requires a complex multiplication, consisting of four real-valued multiplications with subsequent summation. This would lead to a large number of multiplications to be executed, while achieving a sufficient resolution B for the carrier phase values φ_b. The hardware effort would therefore become prohibitive. Applying the CORDIC (coordinate rotation digital computer) algorithm can dramatically reduce the necessary hardware effort to calculate the B rotated test symbols [50; 51]. This algorithm can compute vector rotations simply by summation and shift operations. As for the calculation of the B rotated copies of the input vector intermediate results can be reused for different rotation angles, only $\sum_{b=1}^{\log_2\{B\}} 2^{b+1}$ shift-and-add operations are required to generate the B test symbols. For example to generate $B = 32$ rotated copies of Z_k the CORDIC algorithm requires only 124 shift-and-add operations instead of 124 real valued multiplications and 62 adders.

3.4.4.3.2 Calculation of the distance to the closest constellation point

To determine the closest constellation point $\hat{X}_{k,b}$ the rotated symbols are fed into a decision circuit. The squared distance $|d_{k,b}|^2$ calculated with equation (3.41) can be rewritten as

$$|d_{k,b}|^2 = (\text{Re}[d_{k,b}])^2 + (\text{Im}[d_{k,b}])^2$$
$$= (\text{Re}[Z_k \exp\{j\varphi_b\}] - \text{Re}[\hat{X}_{k,b}])^2 + (\text{Im}[Z_k \exp\{j\varphi_b\}] - \text{Im}[\hat{X}_{k,b}])^2. \quad (3.45)$$

Implementing this formula literally into hardware would result into two multipliers and three adders/subtractors. But a closer examination of (3.41) and (3.45) reveals that the results of the subtractions are relatively small because the distance to the closest constellation point is calculated. Therefore the most significant bits (MSBs) of the absolute value of the subtraction result will always be zero and can be discarded to reduce the hardware effort. As the required accuracy for

$|d_{k,b}|^2$ is also moderate, it can be determined using a look-up table (LUT) or basic logic functions more efficiently than with multipliers.

3.4.4.3.3 Filter function

Highly parallelized systems allow a very resourceful implementation of the summation of $2N_{CR}+1$ consecutive values. The adders can be arranged in a binary tree structure where intermediate results from different modules are reused in neighboring modules. This leads to a moderate hardware effort.

3.4.5 Hardware effort

In this section the hardware effort is estimated for the different carrier recovery algorithms at the example of QPSK. For this purpose the required number of basic functional blocks such as multipliers, adders/subtractors, look-up tables (LUT), comparators and switches is analyzed. Table 3.2 lists the required blocks for each of the different algorithms. Note that a complex multiplication requires 4 real-valued multiplications and 3 adders. A complex addition or subtraction requires 2 real-valued adders/subtractors.

Table 3.2: Required hardware components for different carrier recovery algorithms

Algorithm		Multiplier	LUT	Adder/Subtractor	Comparator	Switch	Reuse of intermediate results
unweighted	V&V, $u=0$	0	3	$4N_{CR}$	0	0	✓
	V&V, $u=2$	0	5	$4N_{CR}$	0	0	✓
	V&V, $u=4$	10	1	$4N_{CR}+4$	0	0	✓
weighted	V&V, $u=0$	$4N_{CR}$	3	$4N_{CR}$	0	0	✗
	V&V, $u=2$	$4N_{CR}$	5	$4N_{CR}$	0	0	✗
	V&V, $u=4$	$4N_{CR}+10$	1	$4N_{CR}+4$	0	0	✗
MLPA		0	1	$4N_{CR}$	$2N_{CR}$	N_{CR}	✓
SMLPA		0	1	$4N_{CR}$	$4N_{CR}$	$2N_{CR}$	✓
QAM		0	B	$\sum_{b=1}^{\log_2\{B\}} 2^{b+1} + 2BN_{CR}$	B	B	✓

The hardware effort for the functional blocks in Table 3.2 reduces from left to right. Thus the most hardware-efficient algorithm is the MLPA. The total component count may be larger than for the unweighted Viterbi & Viterbi (V&V) algorithm with $u = 0$, but the 2 additional look-up tables are more costly than some comparators and switches.

Table 3.2 also reveals the big disadvantage of the weighted V&V algorithm. Not only requires the algorithm a large number of multipliers, additionally due to the weighting almost no intermediate results from the summation process can be reused in other parallel modules. Thus the hardware effort is significantly larger than for the other algorithms except the QAM carrier recovery.

3 Digital signal processing algorithms for coherent optical receivers

The estimated hardware effort for the QAM carrier recovery is roughly B times larger than for the Viterbi & Viterbi or the (S)MLPA algorithm. This may be viewed as a large increase in hardware consumption since B is in the range of 16 to 64. But taking into account that the algorithm is designed for high-order QAM constellations, and that the hardware effort for polarization control or chromatic dispersion (CD) and polarization mode dispersion (PMD) compensation is also much higher than the hardware effort for QPSK carrier recovery, the implementation of the QAM carrier recovery algorithm can still be achieved with reasonable effort.

3.5 Data recovery

When the carrier phase estimate $\hat{\psi}_{IF,k}$ is available the transmitted data can be recovered. The data recovery process is slightly different for QAM constellations with equidistant-phases and square QAM constellations.

3.5.1 Data recovery for QAM constellations with equidistant-phases

For QAM constellations with equidistant-phases the amplitude and phase information can be recovered independently.

The amplitude information n_a can be recovered without carrier recovery. It can be extracted from the magnitude of the received signal by simple threshold decision. The optimum thresholds can be determined from the average power of the signal and depend on the number of amplitude levels in the constellation and the number of constellation points per amplitude level.

The recovered phase information represented by the estimated sector number $\hat{n}_{d,k}$ depends on the recovered carrier. It can be determined with $\hat{\psi}_{IF,k}$ by

$$\hat{n}_{d,k} = \left\lfloor (\varphi_k - \hat{\psi}_{IF,k}) \frac{p}{2\pi} + \frac{1}{2} \right\rfloor, \qquad (3.46)$$

where $\lfloor x \rfloor$ is the biggest integer $\leq x$. The differential sector number $\hat{n}_{d,k}$ introduced in section 2.1.3 is used at this point because $\hat{\psi}_{IF,k}$ has a p-fold ambiguity and thus the absolute phase cannot be unambiguously recovered and differential encoding/decoding is required.

To overcome the ambiguity of $\hat{\psi}_{IF}$ the statistical properties of ψ_{IF} can be exploited. As described in section 2.2.4.3 phase noise can be modeled as a Wiener-Lévy process using equation (2.36). Hence the most likely value for $\hat{\psi}_{IF,k}$ is the one closest to $\hat{\psi}_{IF,k-1}$. However choosing the correct value from the interval $[0, 2\pi)$ requires the availability of $\hat{\psi}_{IF,k-1}$. This violates the constraint "Feasibility of parallel processing" described in section 3.1.1. A solution to this problem is proposed in [52]. The idea is to choose $\hat{\psi}_{IF,k}$ always from the interval $[0, 2\pi/p)$ and to extend the formula (2.6) for differential decoding to

$$\hat{n}_{t,k} = \left(\hat{n}_{d,k} - \hat{n}_{d,k-1} + n_{j,k}\right) \mathrm{mod}\left(\max\{n_t\}+1\right). \quad (3.47)$$

$n_{j,k}$ is referred to as jump number because it indicates when the physical course of $\hat{\psi}_{IF,k}$ crosses the boarders of the interval $[0, 2\pi/p)$ and the actual course of $\hat{\psi}_{IF,k}$ makes a jump $|\hat{\psi}_{IF,k} - \hat{\psi}_{IF,k-1}| > \pi/p$. Therefore $n_{j,k}$ is calculated by

$$n_{j,k} = \begin{cases} 1 & \hat{\psi}_{IF,k} - \hat{\psi}_{IF,k-1} < -\pi/p \\ 0 & |\hat{\psi}_{IF,k} - \hat{\psi}_{IF,k-1}| \leq \pi/p \\ -1 & \hat{\psi}_{IF,k} - \hat{\psi}_{IF,k-1} > \pi/p \end{cases}. \quad (3.48)$$

3.5.2 Data recovery for square QAM constellations

For the data recovery from square QAM constellations the evaluation of amplitude and phase information cannot be separated like for equidistant-phase constellations. Thus no data can be recovered without previous carrier recovery.

At first the differentially encoded quadrant number $n_{d,k}$ is recovered from the square QAM constellation. The decoding process is the same as for the phase information in QAM constellations with equidistant-phases and $p = 4$:

$$\hat{n}_{d,k} = \left\lfloor (\varphi_k - \hat{\psi}_{IF,k})\frac{2}{\pi} + \frac{1}{2} \right\rfloor \quad (3.49)$$

Using this information the recovered symbol can be rotated into the first quadrant of the complex plane by

$$\underline{Y}'_k = \underline{Y}_k e^{-j\left(\hat{\psi}_{IF,k} + \hat{n}_{d,k}\frac{\pi}{2}\right)}. \quad (3.50)$$

The inphase and quadrature numbers $n_{i,k}$ and $n_{q,k}$ can be recovered from $\mathrm{Re}\{\underline{Y}_k'\}$ and $\mathrm{Im}\{\underline{Y}_k'\}$, respectively, using two threshold deciders. The threshold levels can be determined from the average signal power and depend on the number of constellation points.

Finally the quadrant number $\hat{n}_{d,k}$ has to be differentially decoded to obtain $\hat{n}_{t,k}$. The differential decoding process is the same as for QAM constellations with equidistant-phases given by equation (3.47) and $\max\{n_t\} = 3$.

$$\hat{n}_{t,k} = \left(\hat{n}_{d,k} - \hat{n}_{d,k-1} + n_{j,k}\right) \mathrm{mod}\, 4 \quad (3.51)$$

Only the calculation of the jump number $n_{j,k}$ has to be adapted to the carrier recovery process and is calculated with

$$n_{j,k} = \begin{cases} 1 & b_{\min,k} - b_{\min,k-1} < -B/2 \\ 0 & |b_{\min,k} - b_{\min,k-1}| \leq B/2 \\ -1 & b_{\min,k} - b_{\min,k-1} > B/2 \end{cases} \qquad (3.52)$$

3.6 Intermediate frequency control

In a coherent optical receiver with feed-forward carrier recovery a locking of the signal and local oscillator laser frequencies is not required. However the frequencies of the lasers should be sufficiently close to not degrade the carrier recovery process. Therefore either an external LO frequency control or internal intermediate frequency compensation is required.

3.6.1 External LO frequency control

One option to control the intermediate frequency (IF) between the signal and local oscillator is to directly control the LO laser frequency, e.g. by altering a portion of the LO bias current. In this book the method will be referred to external frequency control. As control error signal the jump numbers $n_{j,k}$ defined in equation (3.48) can be used. If the course of the recovered carrier phase continuously deceases the LO frequency is too low. If it continuously increases the LO frequency is too high. These states are indicated by the fact that $n_{j,k} = -1$ occurs more often that $n_{j,k} = 1$ or vice versa. Thus by feeding the $n_{j,k}$ into an external integral controller the LO laser frequency can be locked to the signal laser frequency.

3.6.2 Internal intermediate frequency compensation

Instead of directly controlling the LO laser frequency it is also possible to compensate for the IF in the DSPU. Like for the external LO frequency control the jump numbers $n_{j,k}$ can be used as control error signal. But instead of outputting the $n_{j,k}$ from the DSPU they are integrated internally.

$$\hat{\omega}_{IF,k} = \sum_{i=-\infty}^{k} g_{IF} n_{j,i} \qquad (3.53)$$

$\hat{\omega}_{IF,k}$ is the estimated product $\omega_{IF} T$. As only a coarse compensation of the IF is required, the control gain g_{IF} can be small. The IF is compensated by altering the phase of the received symbols.

$$\begin{bmatrix} \underline{Z}'_{k,x} \\ \underline{Z}'_{k,y} \end{bmatrix} = \begin{bmatrix} \underline{Z}_{k,x} \\ \underline{Z}_{k,y} \end{bmatrix} \exp\left\{-j \sum_{i=-\infty}^{k-\Delta} \hat{\omega}_{IF,i}\right\} = \begin{bmatrix} \underline{Z}_{k,x} \\ \underline{Z}_{k,y} \end{bmatrix} e^{-j(\hat{\omega}_{IF,k-\Delta} + \hat{\omega}_{IF,k-\Delta-1})} \qquad (3.54)$$

Δ is the processing delay of the system.

4 Simulation results

The performance of the polarization control, ISI compensation and carrier recovery algorithms described in chapter 3 is investigated in extensive simulations. The goals are to define the optimum system parameters such as control gain or filter width and to determine the tolerances against various distortions such as AWGN, polarization cross-talk, PMD, phase noise and quantization effects. Especially the simulation results for a QPSK transmission system are required for the implementation of a real-time polarization-multiplexed synchronous QPSK system. All simulations were programmed and executed with MATLAB®.

4.1 QPSK carrier recovery

The feed-forward carrier recovery algorithms for a QAM constellation with equidistant-phases described in section 3.4 are compared in Monte-Carlo simulations of a QPSK transmission system. The QPSK constellation is chosen as a reference because a polarization-multiplexed QPSK system is the target system of the synQPSK project to be realized in a real-time laboratory experiment. The following simulation results were used to determine the main system parameters. Another more general reason for the detailed investigation of the QPSK format is its superior noise immunity that makes it most attractive for long-haul fiber transmission systems. Each data point is based on the simulation of 2,000,000 symbols. The results are compared against the theoretically optimal receiver sensitivity [19] given by

$$\frac{E_S}{N_0} = Q^{-1}[\text{BER}]. \qquad (4.1)$$

E_S/N_0 is the normalized optical signal to noise ratio introduced in section 2.2.3.5, and BER is the target bit error rate. Note that this equation does not take the coding penalty due to differential encoding/decoding into account. Thus the differential coding penalty $F = 2$ for QPSK (Table 2.1) contributes to the sensitivity penalties determined by the simulations.

As another reference also the feed-forward carrier recovery for arbitrary QAM constellations described in section 3.4.4 is analyzed for QPSK. The hardware-effort for this algorithm is much higher than for the other considered algorithms, thus for a practical QPSK implementation the QAM carrier recovery will be of low interest. But QPSK is the only modulation format, where the QAM algorithm performance can be compared against efficient state-of-the-art techniques. For higher-order QAM formats only poorly conceived techniques exist which have either a low practicality or a low phase noise tolerance.

4 Simulation Results

4.1.1 QPSK carrier phase estimator efficiency and mean squared error

In the following the mean squared error (MSE) and the efficiency $e(N_{CR})$ of the different carrier phase estimators are analyzed. The MSE allows quantifying how much the estimated carrier phase $\hat{\psi}$ differs from the true value of ψ and is given by

$$\text{MSE}(\hat{\psi}) = \left\langle (\psi - \hat{\psi})^2 \right\rangle. \tag{4.2}$$

To determine the efficiency of an estimator, its MSE is compared against the Cramér-Rao lower bound (CRLB), which specifies the lowest possible mean squared error achievable by an unbiased estimator [53]. To calculate the CRLB the contribution of AWGN to phase noise is approximated to be Gaussian with the variance $(2E_S/N_0)^{-1}$ [54]. With this approximation, which has a high accuracy for high OSNRs, the CRLB depending on the filter half width N_{CR} is given by

$$\text{CRLB}(N_{CR}) = \frac{1}{2N_{CR}+1} \cdot \frac{N_0}{2E_S}. \tag{4.3}$$

Then the efficiency $e(N_{CR})$ of the carrier phase estimators is defined as

$$e(N_{CR}) = \frac{\text{CRLB}(N_{CR})}{\text{MSE}(\hat{\psi})} = \frac{N_0}{2E_S(2N_{CR}+1)\left\langle(\psi-\hat{\psi})^2\right\rangle} \leq 1. \tag{4.4}$$

As the CRLB is the lowest possible mean squared error, the maximum estimator efficiency is 1.

4.1.1.1 Carrier phase estimator efficiency for $\Delta f \cdot T_S = 0$

Figure 4.1 shows the efficiency $e(N_{CR})$ in the absence of laser-induced phase noise ($\Delta f \cdot T_S = 0$) and two different OSNRs for all carrier recovery algorithms that were presented in section 3.4. Normally for $N_{CR} = 0$ one would expect that $e(0) = 100\%$, because the actual symbol phase is used as carrier phase, i.e. no estimation of the carrier is performed. The reason for $e(0) < 100\%$ is that the approximation of the AWGN-induced phase noise variance is too optimistic, especially for lower OSNRs.

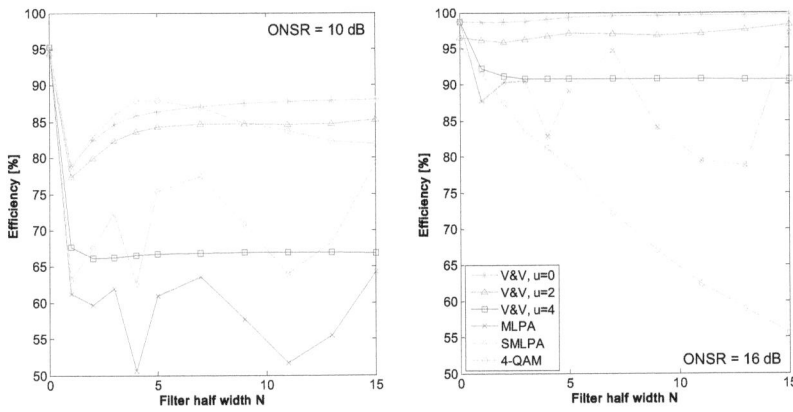

Figure 4.1: QPSK carrier phase estimator efficiency for OSNR = 10 dB (left) and OSNR = 16 dB (right)

The comparison of the different carrier phase estimators shows that the Viterbi & Viterbi (V&V) algorithm with $u = 0$ yields the best performance reaching almost 100% efficiency for OSNR = 16 dB. Increasing u to 2 or 4 degrades the efficiency. This becomes evident if one considers that the amplitude contains no information about the carrier phase ψ_{IF}. As the amplitude of a QPSK signal is constant, amplitude fluctuations are only caused by noise, whose influence can be reduced by considering only the phase of the received symbols for carrier recovery. Note that a differentiation between the Viterbi & Viterbi algorithms with and without weighted averaging is not necessary for $\Delta f \cdot T_S = 0$, because all w_i have the same value and thus both filters are identical.

Comparing the MLPA algorithm against the other approaches, for OSNR = 10 its performance is inferior. But it can be significantly improved by adding reliability information to each intermediate result, referred to as the SMLPA algorithm. The effect that the two algorithms do not generate a smooth curve is caused by the filter structures that vary for each value of N_{CR} (compare Table 3.1). The reason that for OSNR = 16 the two algorithms yield exactly the same performance is due to the fact that for high OSNRs virtually all intermediate results are reliable and therefore the output of both algorithms is the same.

The 4-QAM carrier recovery achieves an efficiency comparable to the V&V algorithm with $u = 0$, admittedly only for OSNR = 10 and short filter lengths. And for OSNR = 16 dB the efficiency even degrades dramatically for larger filters. This result is not surprising if one recalls that the QAM carrier recovery algorithm has an inherent phase quantization, which is set to $\log_2\{B\} = 5$ in this simulation. This limits the achievable mean squared error of the estimation process and thus degrades the efficiency, especially for low OSNRs and large filters, where the CRLB is low.

4 Simulation Results

4.1.1.2 Carrier phase estimator mean squared error for $\Delta f \cdot T_S > 0$

In the presence of laser-induced phase noise ($\Delta f \cdot T_S > 0$) the analysis of the estimator efficiency yields only minor information, because the CRLB is only valid for the estimation of constant values. But this requirement is not any more fulfilled if laser phase noise is considered next to AWGN. Therefore in this section the mean squared errors of the estimator outputs are depicted. The CRLB is plotted as a reference curve.

Figure 4.2 shows the mean squared error of the different carrier recovery algorithms for OSNR = 10 dB and OSNR = 16 dB. As the linewidth of DFB lasers employed in commercial optical transmission systems is usually in the range of 1 MHz to 10 MHz and the symbol rates range from 10 Gbaud to 40 Gbaud, the carrier recovery should be able to tolerate linewidth-times-symbol-duration products as high as $\Delta f \cdot T_S = 10^{-3}$. Therefore the considered linewidth-times-symbol-duration products are $\Delta f \cdot T_S = 10^{-4}$ (top left), $\Delta f \cdot T_S = 4 \cdot 10^{-4}$ (top right) and $\Delta f \cdot T_S = 10^{-3}$ (bottom).

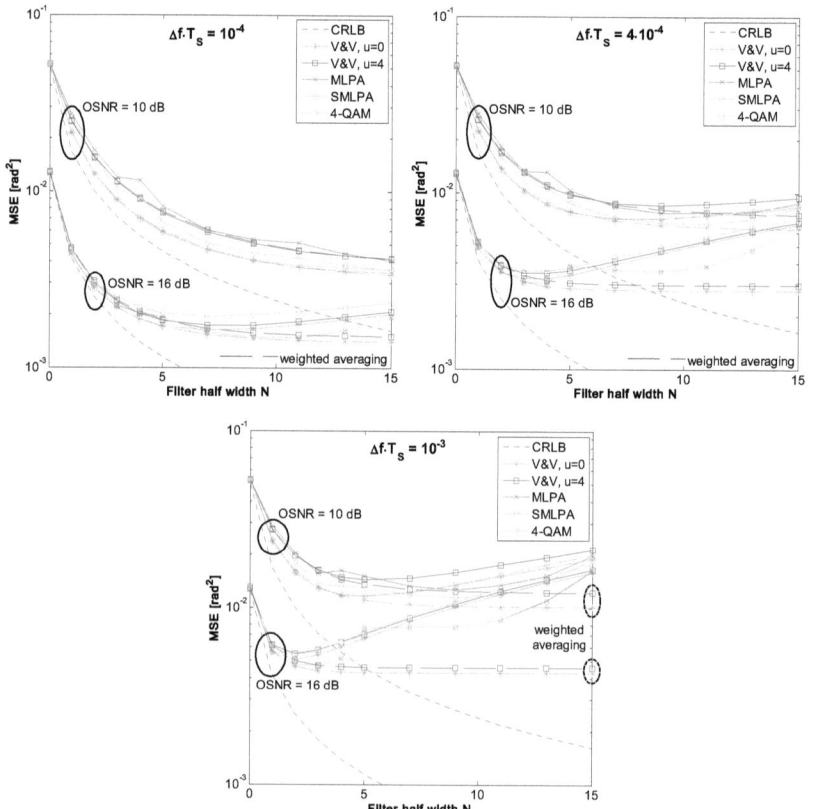

Figure 4.2: Carrier phase estimator mean squared error for different values of $\Delta f \cdot T_S$

Note that the Viterbi & Viterbi algorithm with $u = 2$ is not considered any more, because its performance is inferior to the corresponding algorithm with $u = 0$ while the required computational effort for it is higher.

The results for the MSE are similar to the ones obtained for the estimator efficiency for $\Delta f \cdot T_S = 0$. The best performance is also achieved using the Viterbi & Viterbi algorithm with $u = 0$ – but only if weighted averaging is applied. Especially for higher phase noise values the SMLPA algorithm often outperforms the Viterbi & Viterbi algorithm without weighted filter inputs. This can be explained by the fact that depending on the filter structure the SMLPA filter also weights its input values. This is advantageous in the presence of phase noise. As the QAM carrier phase estimator does not have this advantage it is not surprising that its MSE is very similar to the one of the unweighted Viterbi & Viterbi algorithm.

Figure 4.2 also unveils that most algorithms have an optimum filter width that varies for different values of $\Delta f \cdot T_S$. The optimum values for N_{CR} are between 2 and 6 for $\Delta f \cdot T_S = 10^{-3}$ and increase for lower linewidth-times-symbol-duration products. An exception to this is the Viterbi & Viterbi carrier recovery with weighted averaging. As its weighting coefficients are adaptively optimized to the ratio of laser phase noise to AWGN-induced phase noise the value for the optimum filter half width is $N_{CR} \to \infty$. However although the MSE for the Viterbi & Viterbi algorithm does not degrade, but it only improves marginally if N_{CR} becomes larger than the optimum filter half widths of the other approaches. The reason is that the outer weights of the filter tend to zero.

4.1.2 QPSK phase noise tolerance

In order to evaluate the effect of the different algorithm efficiencies and MSEs on the receiver sensitivity of a coherent optical receiver, the BER for the different carrier recovery algorithms against the OSNR is investigated. The same values are used for the linewidth-times-symbol-duration products as in section 4.1.1. $N_{CR} = 4$ and $N_{CR} = 9$ are the considered filter half widths.

4.1.2.1 Viterbi & Viterbi algorithm performance

Figure 4.3 shows the BER against the OSNR for the original Viterbi & Viterbi feed-forward carrier recovery algorithm described in section 3.4.1 for $u \in \{0, 4\}$.

4 Simulation Results

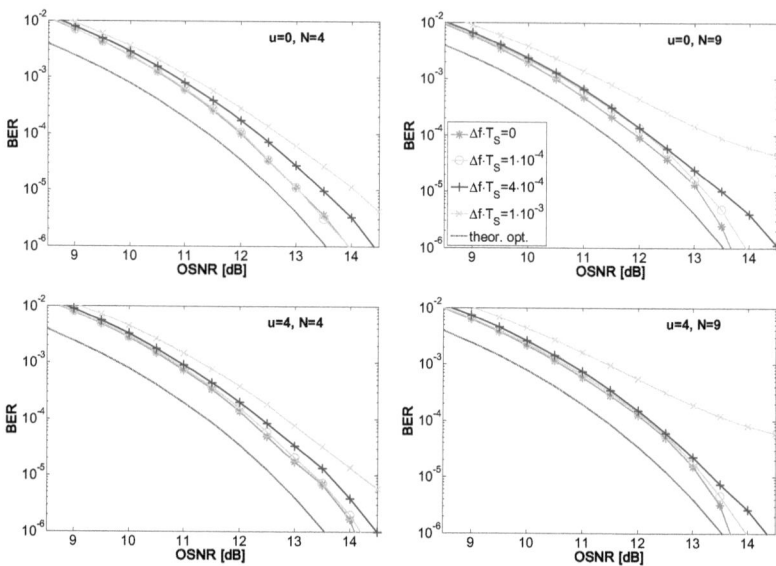

Figure 4.3: OSNR vs. BER for Viterbi & Viterbi carrier recovery
and different linewidth-times-symbol-duration products $\Delta f \cdot T_S$.

The comparison of the receiver sensitivity for $u = 0$ and $u = 4$ confirms the results from section 4.1.1 that the neglect of the symbol amplitude ($u = 0$) is beneficial for the carrier recovery performance. Looking at the different values of the filter half width N_{CR}, then no general decision can be made for an optimum value. For low phase noise a large filter is beneficial to reduce the negative effects of AWGN. But for high values of $\Delta f \cdot T_S$ the symbols at the edges of the filter carry only little information about the actual carrier phase to be recovered. Therefore for the unweighted Viterbi & Viterbi algorithm a narrower filter becomes advantageous.

As expected from the results in section 4.1.1 the simulation results for the weighted Viterbi & Viterbi algorithm depicted in Figure 4.4 do not contain such a performance degradation for high values of $\Delta f \cdot T_S$ and large filters. Especially for $u = 0$ and $N_{CR} = 9$ the benefit of the weighted filtering becomes obvious. Due to the adaptation of the weighting coefficients to the ratio of laser phase noise to the angular portion of AWGN ($\sigma_\Delta^2 / \sigma_{n'}^2$) no penalty is observed compared to the results for $N_{CR} = 4$, even for $\Delta f \cdot T_S = 10^{-3}$.

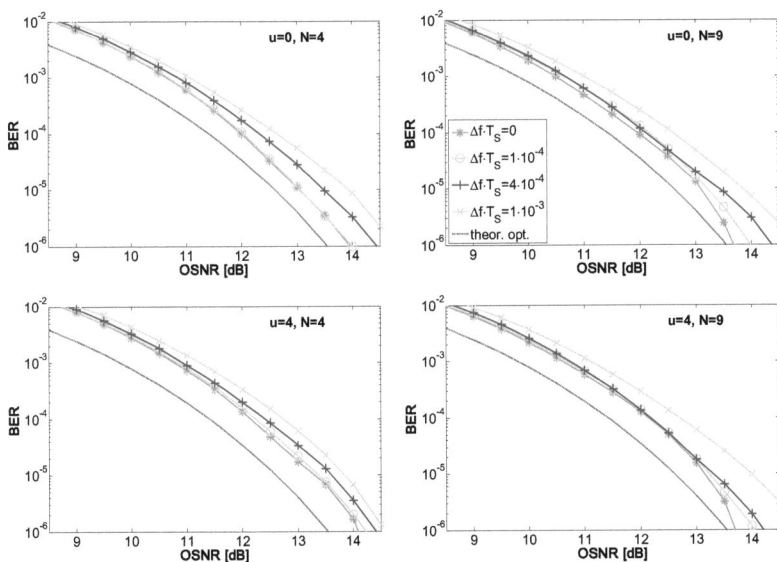

Figure 4.4: OSNR vs. BER for weighted Viterbi & Viterbi carrier recovery
and different linewidth-times-symbol-duration products.

Figure 4.5 compares the sensitivity penalties at BER = 10^{-3} both for the unweighted (left) and weighted (right) Viterbi & Viterbi carrier recovery algorithms. For $\Delta f \cdot T_S < 10^{-4}$ the sensitivity improves only marginally. The reason for the residual sensitivity penalty for $\Delta f \cdot T_S \rightarrow 0$ is the differential coding penalty explained in section 2.1.3 and an additional implementation penalty depending on the efficiency of the carrier recovery process.

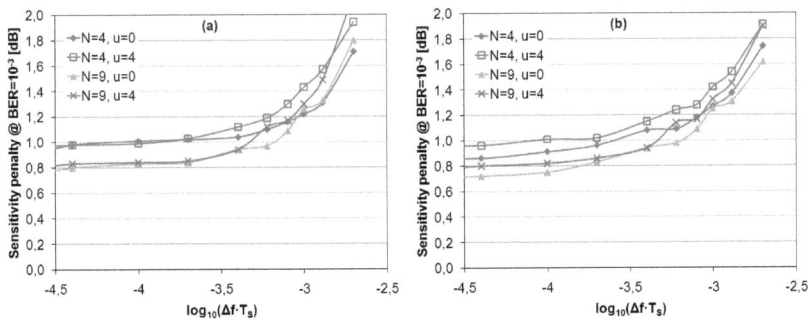

Figure 4.5: Sensitivity penalties at BER = 10^{-3} against $\Delta f \cdot T_S$ for unweighted (a)
and weighted (b) Viterbi & Viterbi carrier recovery

4 Simulation Results

4.1.2.2 (S)MLPA algorithm performance

In this subsection the phase noise tolerance of the (S)MLPA algorithm presented in section 3.4.3 is evaluated. Both the original MLPA algorithm, also referred to as barycenter algorithm, and the SMLPA algorithm which uses additional reliability information about intermediate results are considered. Figure 4.6 depicts the corresponding BERs against OSNR.

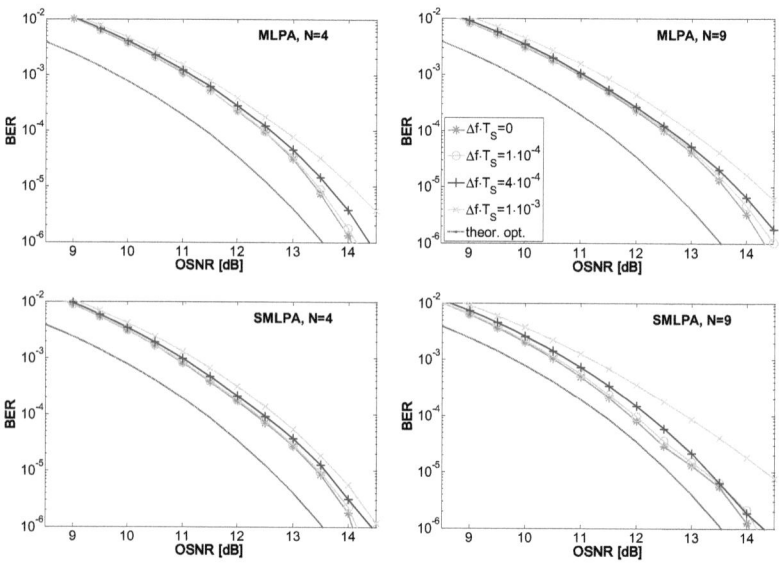

Figure 4.6: OSNR vs. BER for (S)MLPA carrier recovery and different linewidth-times-symbol-duration products

The results show that for the MLPA algorithm a larger filter gives only little benefits. For low OSNRs the improvement due to the larger filter is marginal, and for high OSNRs the sensitivity even degrades for $\Delta f \cdot T_S = 0$. This is due to the fact that the carrier phase angles are averaged pairwise. As for each averaging step one of three possible solutions has to be selected, the error-susceptibility increases for larger filter widths, even for low linewidth-times-symbol-duration products.

This drawback is overcome by adding reliability information to each intermediate result. The simulation results for SMLPA in the bottom row of Figure 4.6 confirm a sensitivity improvement, especially for low OSNRs and larger filters. Only for $N_{CR} = 9$ and $\Delta f \cdot T_S = 10^{-3}$ the phase noise tolerance is slightly degraded. This can also be seen in Figure 4.7, which exemplifies the different sensitivity penalties at BER = 10^{-3}.

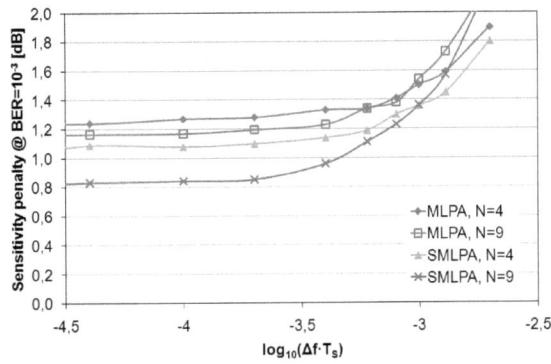

Figure 4.7: Sensitivity penalties at BER = 10^{-3} against $\Delta f \cdot T_S$ for (S)MLPA carrier recovery

4.1.2.3 4-QAM carrier recovery performance

Figure 4.8 shows the simulation results for 4-QAM carrier recovery with a carrier phase resolution of $\log_2\{B\} = 5$. The performance for low linewidth-times-symbol-duration products $\Delta f \cdot T_S \leq 4 \cdot 10^{-4}$ is comparable to the other approaches. But for $\Delta f \cdot T_S = 10^{-3}$ the disadvantage of constant weights for the filter inputs becomes visible, similar to the results for the unweighted Viterbi & Viterbi algorithm in section 4.1.2.1. The sensitivity disproportionately reduces as phase noise increases, especially for larger filters. This effect becomes even more obvious in Figure 4.9, which shows the sensitivity penalties for 4-QAM carrier recovery at a BER of 10^{-3}.

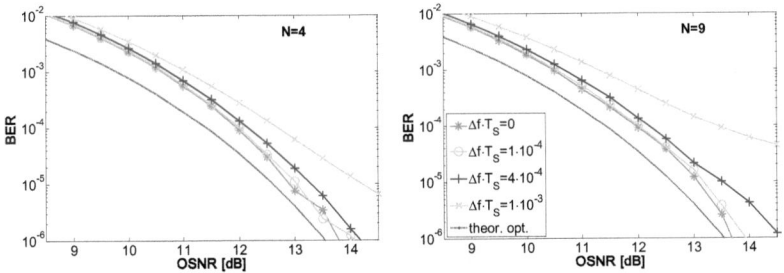

Figure 4.8: OSNR vs. BER for 4-QAM carrier recovery and different linewidth-times-symbol-duration products.

4 Simulation Results

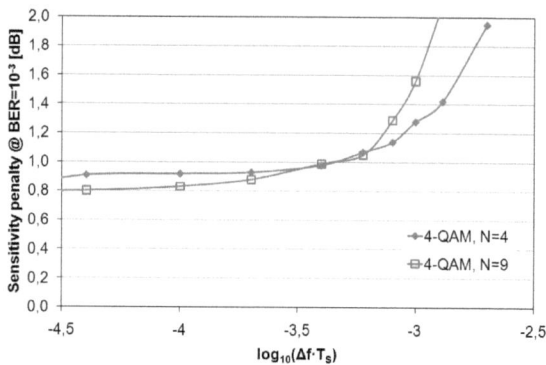

Figure 4.9: Sensitivity penalties at BER = 10^{-3} against $\Delta f \cdot T_S$ for 4-QAM carrier recovery

4.1.2.4 Summary

A comparison between all considered QPSK carrier recovery concepts unveils that the best performance, i.e. the highest receiver sensitivity is achieved by applying the weighted Viterbi & Viterbi algorithm with $u = 0$ and $N_{CR} = 9$. Its minimum penalty compared to the theoretical optimum is 0.7 dB and it has a high tolerance against phase noise. The barycenter algorithm extended by reliability information (SMLPA) achieves roughly the same phase noise tolerance, but its minimum sensitivity penalty exceeds the penalty of the weighted Viterbi & Viterbi algorithm by 0.1 dB. Finally the 4-QAM carrier recovery achieves the same receiver sensitivity as the barycenter algorithm, even though it contains an inherent quantization of the phase, which is not considered yet for the other algorithms. But due to its unweighted filter inputs its phase noise tolerance is inferior.

All in all the differences in receiver sensitivity due to the different carrier recovery algorithms are small. Therefore it is difficult to give a final recommendation, which algorithm should be employed in a real system implementation. If the strategic decision is to develop a receiver with ultimate performance, then the Viterbi & Viterbi algorithm with adaptive weighted averaging should be applied. However, the SMLPA algorithm offers a high potential to reduce the number of required logic gates in a hardware implementation at only the minor cost of slightly reduced receiver sensitivity.

4.1.2.5 Common carrier in a polarization-multiplexed QPSK receiver

In a polarization-multiplexed QPSK receiver a common carrier can be recovered from the data of both polarizations. Figure 4.10 shows a comparison of the sensitivity penalties for the Viterbi & Viterbi algorithm without weighting, the SMLPA algorithm and the QAM carrier recovery for a single-polarization system with $N_{CR} = 9$ and a polarization-multiplexed system with $N_{CR} = 4$. Thus almost the same amount of 19 and 18 symbols, respectively, is used to recover the carrier.

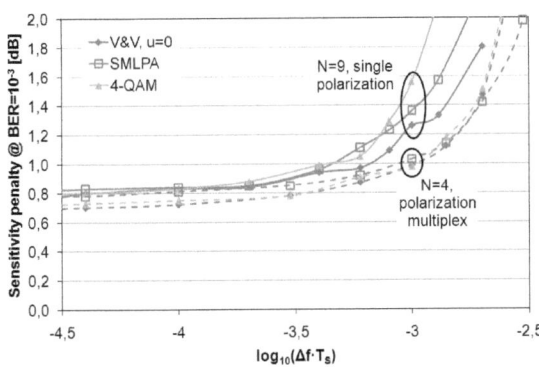

Figure 4.10: Phase noise tolerance for different carrier recovery algorithms using either the data from a single-polarization or from both polarizations

For $\Delta f \cdot T_S = 0$ the two systems using the same carrier recovery algorithm also obtain the same receiver sensitivity. As the estimation process is based on the same number of symbols this result is not surprising. But the systems with joined carrier recovery tolerate roughly twice as much phase noise as the respective algorithms using the data from only one polarization. The reason for this better tolerance is that the algorithms with joined carrier recovery require only half the filter width to recover the carrier phase from the same amount of data. Considering that the symbols that are temporally closer to the symbol to be decoded carry more information about its carrier phase than symbols that are temporally further away, the higher accuracy of the joined carrier recovery process becomes evident.

4.1.3 QPSK analog-to-digital converter resolution

In this section the influence of the analog-to-digital converter (ADC) resolution on the receiver sensitivity is analyzed. As the sensitivity penalty only marginally depends on the employed carrier recovery concept, the considered algorithms are reduced to the Viterbi & Viterbi (V&V) algorithm with $u = 0$, the SMLPA and the 4-QAM carrier recovery algorithms. As the simulations are conducted with $\Delta f \cdot T_S = 0$ a differentiation between V&V algorithms without and with weighted averaging is not necessary. The filter half width is set to $N_{CR} = 9$.

Figure 4.11 depicts the sensitivity penalty of the receiver due to the quantization of the input samples by the ADCs. It is remarkable that the sensitivity penalty for the QAM carrier recovery only slightly degrades for ADC resolutions down to 4 bit and is even below the penalty of the ideal receiver. This can be explained by the fact that the QAM carrier recovery does not calculate the carrier phase as a function of the input samples, but selects the carrier phase angle that allows the most reliable decoding. Therefore the dependence on the accuracy of the input samples is reduced. For the other carrier recovery algorithms (V&V and SMLPA), that determine the carrier phase directly from the input samples, sensitivity starts degrading for ADC resolutions below 5 bit.

4 Simulation Results

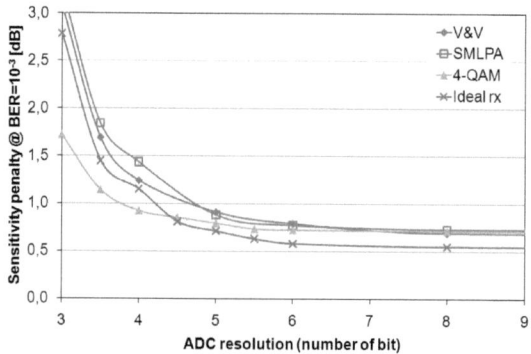

Figure 4.11: Sensitivity penalty vs. analog-to-digital converter resolution for different QPSK carrier recovery algorithms

4.1.4 QPSK phase resolution

An important internal resolution for the different QPSK carrier recoveries is the required precision of the symbol phase. All algorithms require the calculation of the phase from the input symbol. But the received symbol is given by its real and imaginary part. Hence a transformation into the angular domain is required. As the calculation of the argument of a complex number is very complex to realize literally in hardware, usually the results for all possible inputs are stored in a look-up table (LUT). Therefore the required precision of these results is an important parameter for the digital implementation of the algorithms.

Figure 4.12 shows the sensitivity penalty due to the phase quantization. Even down to a 3 bit resolution of the symbol phase the data can still be recovered from the received symbol with only low performance degradation. However to avoid any penalty a 5 or 6 bit resolution should be chosen.

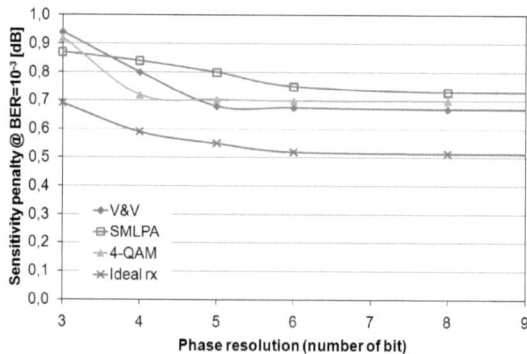

Figure 4.12: Sensitivity penalty vs. phase resolution for different QPSK carrier recovery algorithms

4.2 QAM carrier recovery

In this section the performance of the QAM carrier recovery algorithm proposed in section 3.4.4 is evaluated. The considered constellations are 4-QAM (QPSK), 16-QAM, 64-QAM and 256-QAM. The simulations are limited to square constellations because they are easy to generate [17; 55] and have a high noise immunity in AWGN-dominated transmission systems [21]. The filter width is always set to $N_{CR} = 9$, and each data point is based on the simulation of 200,000 symbols. The results are compared against the theoretically achievable sensitivity calculated with the following formula [19]:

$$\frac{E_S}{N_0} = \frac{M-1}{3} \left(Q^{-1} \left[\frac{\log_2\{M\}}{2} \left(1 - \frac{1}{\sqrt{M}}\right)^{-1} \left(1 - \sqrt{1-\text{BER}}\right) \right] \right)^2 \quad (4.5)$$

E_S/N_0 is the normalized optical signal to noise ratio (OSNR), M is the number of constellation points and BER is the target bit error rate.

4.2.1 Square QAM phase angle resolution

A crucial quantity for the proposed algorithm is the required number B of the test phase values φ_b. If the required resolution is too large, the realization in hardware becomes unfeasible. Figure 4.13 shows the sensitivity penalty at the bit error rate 10^{-3} for 4-QAM and 16-QAM. The proposed algorithm and a receiver with ideal carrier recovery were simulated with different resolutions for the carrier phase. Ideal carrier recovery means that the receiver knows the exact carrier phase (which is only realizable in simulation) and therefore the sensitivity penalty is only caused by differential quadrant encoding and quantization effects.

4-QAM attains a minimum penalty of 0.5 dB for the ideal receiver and 0.7 dB for the proposed algorithm. The penalty difference of 0.2 dB is thus the implementation-induced penalty. For 16-QAM the minimum penalties decreases (0.4 dB for the ideal receiver, 0.6 dB for the proposed algorithm), because only 2 out of 4 transmitted bits are differentially encoded. For all 4 receivers it can be seen that almost no additional penalty is induced due to the quantization of the carrier phase, provided that $\log_2\{B\} \geq 5$. Therefore in all following simulations for 4-QAM and 16-QAM B is set to 32.

4 Simulation Results

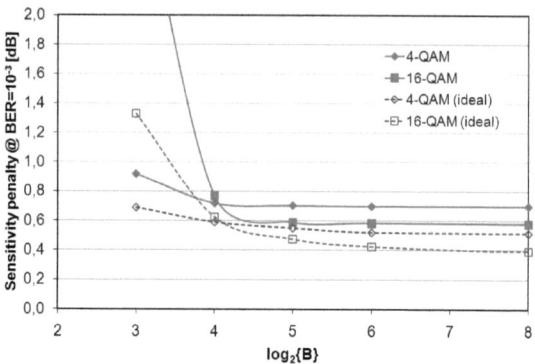

Figure 4.13: Sensitivity penalty for different numbers of test phase values φ_b for 4-QAM and 16-QAM

Figure 4.14 shows the same simulations for 64-QAM and 256-QAM. The minimum penalty for 64-QAM is 0.3 dB with ideal carrier recovery and 0.5 dB using the proposed algorithm. For 256-QAM the respective values are 0.35 dB and 0.55 dB. For both constellations the penalty due to the quantization of the carrier phase is tolerable only if $\log_2\{B\} \geq 6$. The number of test phase values for 64-QAM and 256-QAM is therefore chosen to be $B = 64$ in all subsequent simulations.

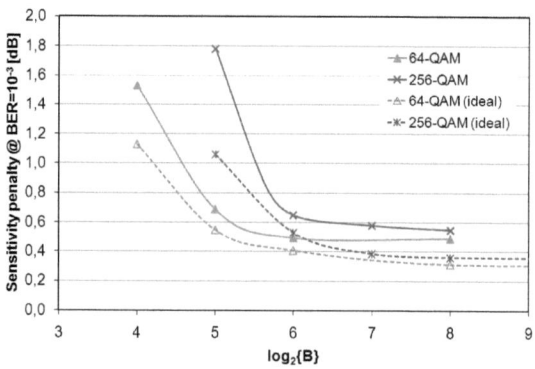

Figure 4.14: Sensitivity penalty for different numbers of test phase values φ_b for 64-QAM and 256-QAM

4.2.2 Square QAM phase estimator efficiency

In the following the efficiency of the QAM phase estimator is analyzed. It is given by the ratio of the CRLB to the mean square error of the estimator. The formula is presented in equation (4.3) of section 4.1.1. As the CRLB is independent of the estimator structure its calculation does not need to be modified for higher-order QAM constellations.

Figure 4.15 shows for 4-QAM and 16-QAM the mean squared error of the phase estimator together with the theoretical optimum expressed by the CRLB (top row) and the resulting estimator efficiency $e(N_{CR})$ (bottom row) for the proposed carrier recovery concept. Figure 4.16 depicts the same information for 64-QAM and 256-QAM.

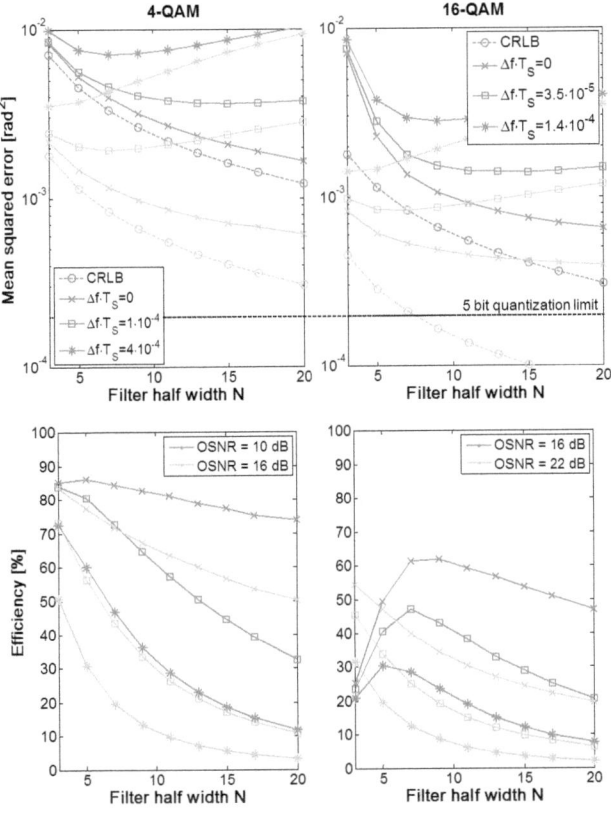

Figure 4.15: Phase estimator mean squared error and efficiency $e(N_{CR})$ vs. filter half width N_{CR} for square 4-QAM (left) and square 16-QAM (right) constellations with $\log_2\{B\} = 5$
(The legends are valid for both figures of a column)

4 Simulation Results

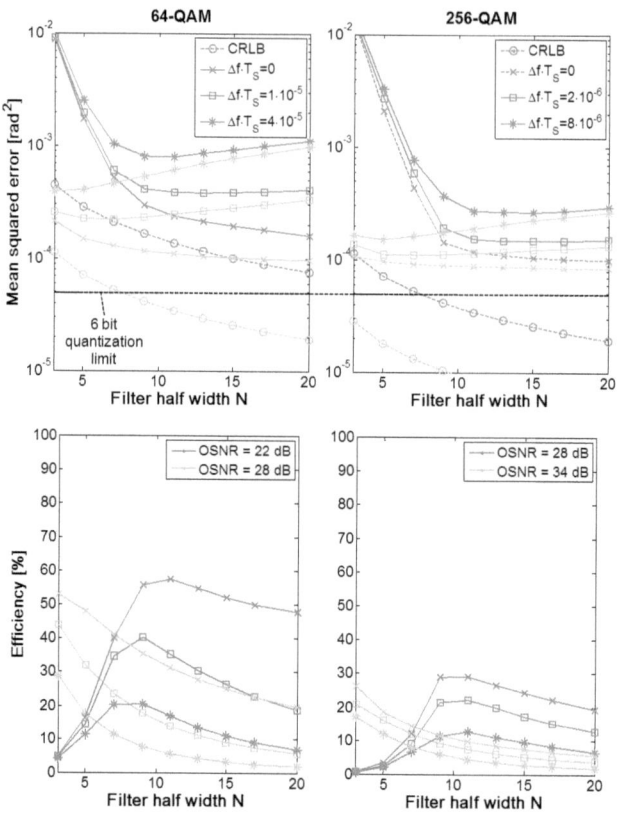

Figure 4.16: Phase estimator mean squared error and efficiency $e(N_{CR})$ vs. filter half width N_{CR} for square 64-QAM (left) and square 256-QAM (right) constellations with $\log_2\{B\} = 6$
(The legends are valid for both figures of a column)

For all considered constellations in the absence of phase noise the mean squared error $\langle(\psi - \hat{\psi})^2\rangle$ continuously decreases for larger values of N_{CR}. In contrast if phase noise is present a global minimum emerges that depends on the linewidth-times-symbol-duration product and the OSNR. It can be seen that $N_{CR} = 9$, which was selected for simulations, induces a mean squared error that is always close to this minimum, especially for lower OSNRs. In principle by optimizing N_{CR} for each parameter set {OSNR, $\Delta f \cdot T_S$, M} the performance of the receiver could have been improved. But as this is not practical in real systems this optimization was omitted in the simulations.

A result comparison among the different QAM constellations makes it apparent that the maximum achievable efficiency reduces from ~85% for 4-QAM to ~60% for 16-QAM and 64-QAM to finally around 30% for 256-QAM. This reduction is mainly caused by the quantization of the carrier phase that limits the minimum achievable mean squared error of the estimator. For $\log_2\{B\} = 5$ (4-QAM,

16-QAM) and $\log_2\{B\} = 6$ (64-QAM, 256-QAM) the minimum mean squared errors are $2 \cdot 10^{-4}$ rad^2 and $5 \cdot 10^{-5}$ rad^2, respectively. For 4-QAM the CRLB is well above the quantization limit, thus quantization effects can mostly be neglected. For 16-QAM and 64-QAM the CRLB is in the range of the quantization limit. Hence the estimator efficiency is already reduced, especially for low OSNR. If the resolution of the carrier phase for 16-QAM is increased to 6 bit, the estimator efficiency improves to ~80% (Figure 4.17). Finally for 256-QAM the CRLB is mostly below the 6 bit quantization limit, which of course severely degrades the estimator efficiency.

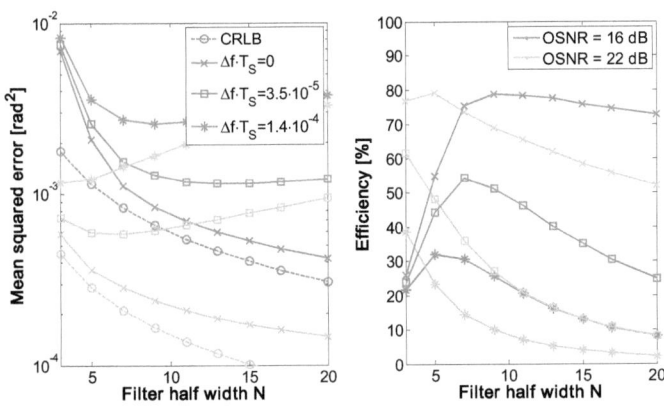

Figure 4.17: Phase estimator mean squared error (left) and efficiency (right) vs. filter half width N_{CR} for a square 16-QAM constellation and $\log_2\{B\} = 6$

(The legends are valid for both figures)

Increasing the carrier phase resolution could improve the estimator efficiency, but at the price of an increased hardware effort.

Figure 4.18 to Figure 4.21 show the considered square QAM constellations at the receiver before and after carrier recovery. They verify that the selected carrier phase resolutions are sufficient to reliably recover the investigated constellations.

4 Simulation Results

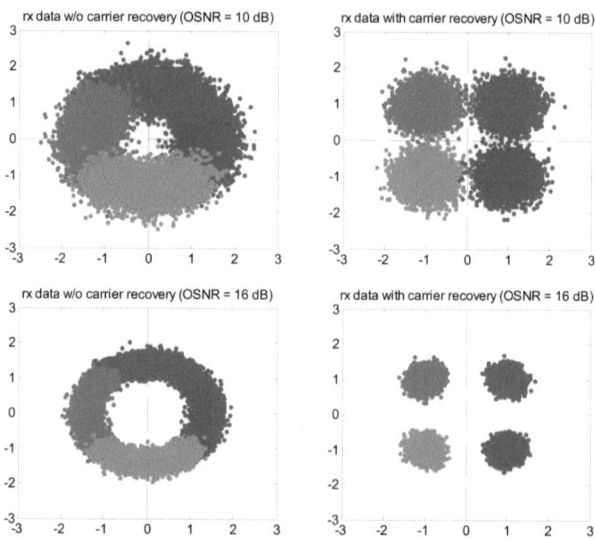

Figure 4.18: 4-QAM constellation diagram at the receiver before and after carrier recovery for $\Delta f \cdot T_S = 4 \cdot 10^{-4}$ ($\log_2\{B\} = 5$, $N_{CR} = 9$)

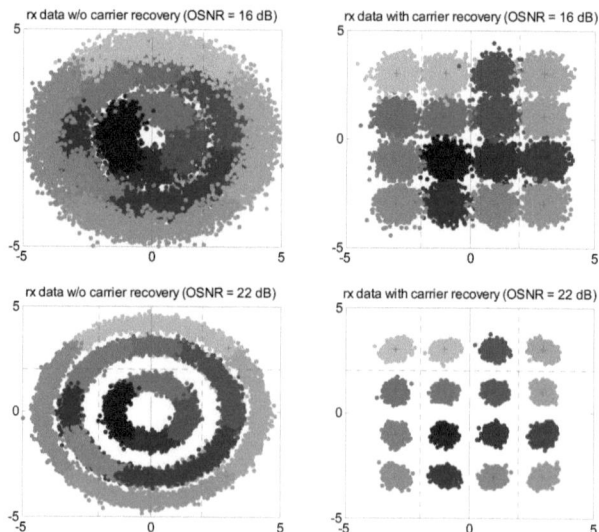

Figure 4.19: 16-QAM constellation diagram at the receiver before and after carrier recovery for $\Delta f \cdot T_S = 1.4 \cdot 10^{-4}$ ($\log_2\{B\} = 5$, $N_{CR} = 9$)

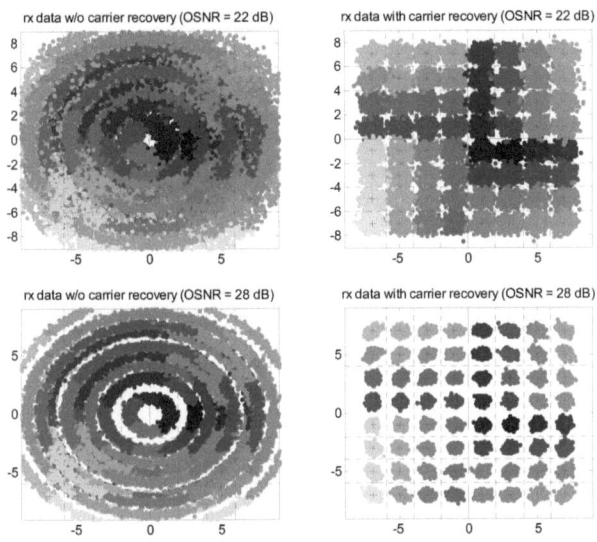

Figure 4.20: 64-QAM constellation diagram at the receiver before and after carrier recovery for $\Delta f \cdot T_S = 4 \cdot 10^{-5}$ ($\log_2\{B\} = 6$, $N_{CR} = 9$)

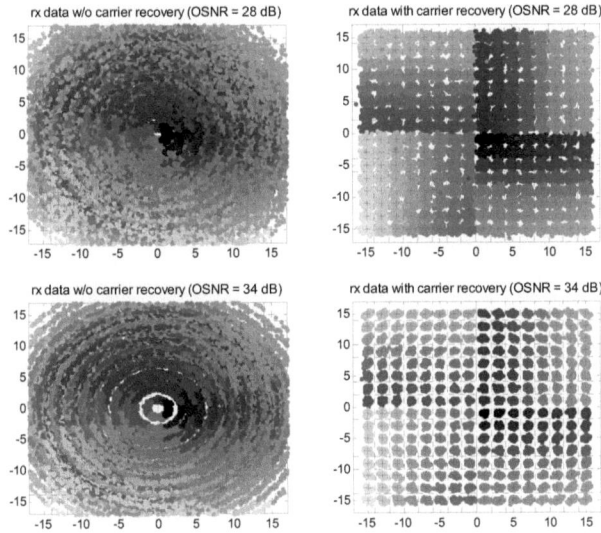

Figure 4.21: 256-QAM constellation diagram at the receiver before and after carrier recovery for $\Delta f \cdot T_S = 8 \cdot 10^{-6}$ ($\log_2\{B\} = 6$, $N_{CR} = 9$)

Another peculiarity in Figure 4.15 and Figure 4.16 is that the efficiency for the higher-order QAM constellations tends to go to zero for short filter lengths and low OSNR. The reason for this effect

4 Simulation Results

becomes obvious if one recalls the carrier recovery algorithm: The squared distance of the received symbol to the closest constellation point is calculated for different carrier phase angles. Looking at the developing of $|d_{k,b}|^2$ and $s_{k,b}$ over the different φ_b shown in Figure 4.22, the algorithm produces for 4-QAM one distinct minimum already for $N_{CR} = 0$. In contrast for higher-order QAM constellations several local minima emerge. Therefore larger filters are required to identify the global minimum, especially for lower OSNRs where the signal is strongly corrupted by noise.

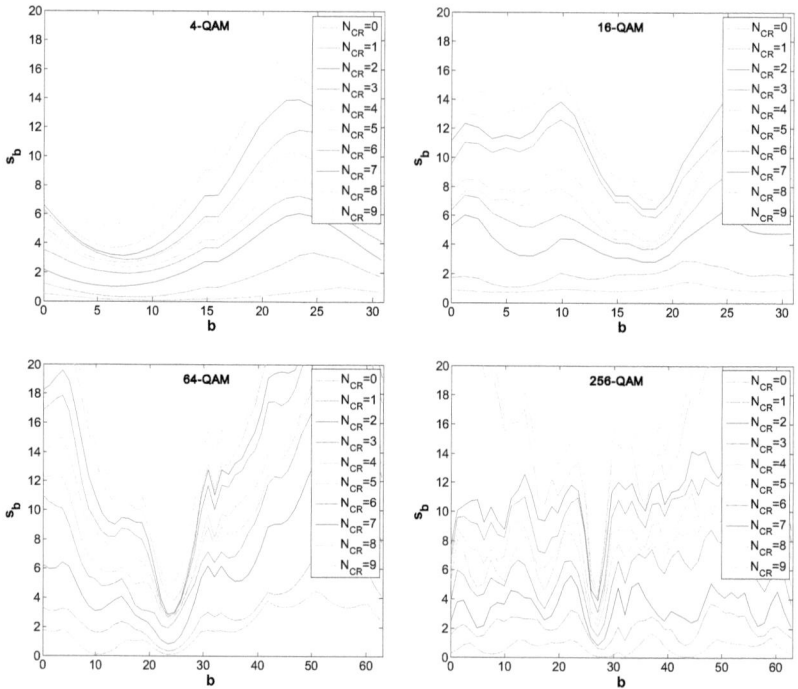

Figure 4.22: Squared distance sum s_b of the b-th parallel block (with test carrier phase angle φ_b) for different filter half widths and different square QAM constellations

4.2.3 Square QAM phase noise tolerance

Another important property of a carrier recovery in a coherent receiver is its tolerance against phase noise. Today's commercial transmission systems usually employ DFB lasers, because they are cost-efficient and have a small footprint. The linewidth of such lasers is in the range of 100 kHz $< \Delta f_{DFB} <$ 10 MHz. Assuming a symbol rate of 10 Gbaud the tolerable linewidth-times-symbol-duration product must be $10^{-5} < \Delta f \cdot T_S < 10^{-3}$.

Figure 4.23 shows the sensitivity penalty of the proposed carrier recovery algorithm against the linewidth-times-symbol-duration product and compares single-polarization carrier recovery with $N_{CR} = 9$ against a joined carrier recovery in a polarization-multiplexed receiver with $N_{CR} = 4$. If no phase noise is present both approaches achieve the same receiver sensitivity. But similar as for the QPSK carrier recovery the phase noise tolerance of all QAM receivers can be roughly doubled by applying a joined carrier recovery for both polarizations. Table 4.1 and Table 4.2 summarize the maximum tolerable linewidths for a 10 Gbaud and a 100GbE system, respectively, for the different square QAM constellations and a sensitivity penalty of 1 dB at BER = 10^{-3}.

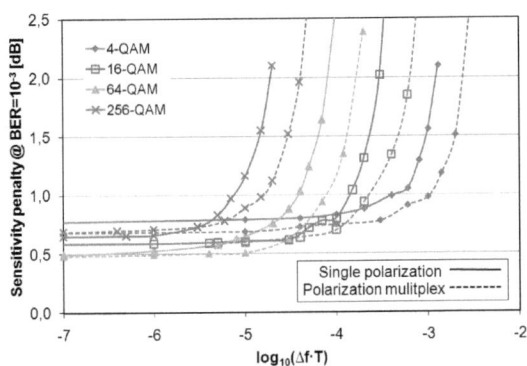

Figure 4.23: Receiver tolerance against phase noise for different square QAM constellations

Table 4.1: Maximum tolerable linewidth for 10 Gbaud systems with different square QAM constellations

Square constellation	Max. tolerable $\Delta f \cdot T_S$ for 1 dB penalty @ BER = 10^{-3}	Max. tolerable Δf for T_S = 10 Gbaud
4-QAM	$4 \cdot 10^{-4}$ ($1 \cdot 10^{-3}$)	4 MHz (10 MHz)
16-QAM	$1.4 \cdot 10^{-4}$ ($2.1 \cdot 10^{-4}$)	1.4 MHz (2.1 MHz)
64-QAM	$4 \cdot 10^{-5}$ ($8 \cdot 10^{-5}$)	400 kHz (800 MHz)
256-QAM	$8 \cdot 10^{-6}$ ($1.5 \cdot 10^{-5}$)	80 kHz (150 kHz)

(The numbers in brackets correspond to a joined carrier recovery in a polarization-multiplexed receiver)

4 Simulation Results

Table 4.2: Required symbol rate and maximum tolerable linewidth to realize a 100GbE (112 Gb/s) system with different square QAM constellations

Square constellation	Bits per symbol	Required symbol rate for 112 Gb/s	Max. tolerable Δf for a 100GbE system
4-QAM	2 (4)	56 Gbaud (28 Gbaud)	22.4 MHz (28 MHz)
16-QAM	4 (8)	28 Gbaud (14 Gbaud)	3.9 MHz (2.9 MHz)
64-QAM	6 (12)	18.67 Gbaud (9.33 Gbaud)	750 kHz (750 kHz)
256-QAM	8 (16)	14 Gbaud (7 Gbaud)	110 kHz (105 kHz)

(The numbers in brackets correspond to a polarization-multiplexed transmission system with joined carrier recovery for both polarization channels at the receiver)

In order to evaluate the phase noise tolerance of the algorithm also for lower BER rates, additional long term simulations have been executed for single-polarization carrier recovery and selected values of $\Delta f \cdot T_S$, simulating 2,000,000 symbols per data point (Figure 4.24). Note that for BERs below 10^{-5} the results become inaccurate due to the low number of errors that occurred during simulation. The theoretical optimum is calculated by inverting equation (4.5):

$$\mathrm{BER} = 1 - \left(1 - \frac{2}{\log_2\{M\}}\left(1 - \frac{1}{\sqrt{M}}\right)\mathrm{Q}\left[\sqrt{\frac{3}{M-1}\frac{E_S}{N_0}}\right]\right)^2 \quad (4.6)$$

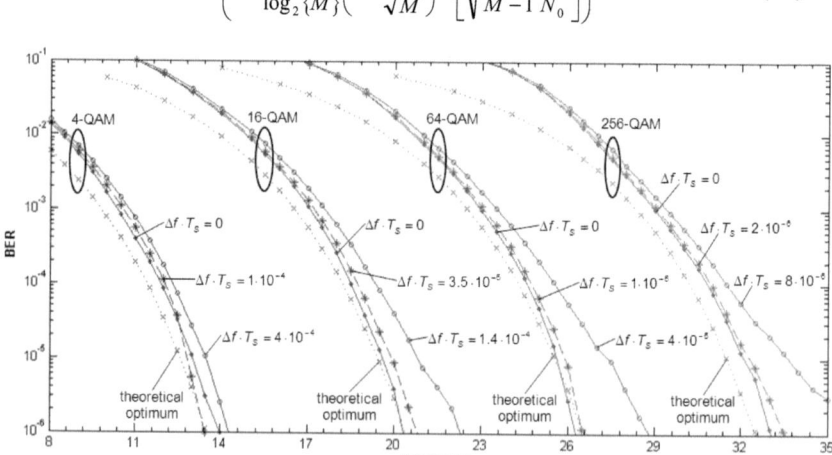

Figure 4.24: Impact of different linewidth-times-symbol-duration products on the receiver sensitivity of coherent QAM receivers

It can be seen that for the values of $\Delta f \cdot T_S$ causing 1 dB of penalty at a BER of 10^{-3}, the penalty increases for lower BERs, especially for higher-order QAM constellations. But if these values are reduced to one quarter the penalty stays almost constant even for BER rates down to 10^{-5}.

4.2.4 Square QAM analog-to-digital converter resolution

Another major obstacle to realize real-time coherent transmission systems with digital carrier recovery is based on the required bandwidth and resolution of the analog-to-digital converter (ADC). Figure 4.25 shows the effect of the ADC resolution on the receiver sensitivity for the considered QAM constellations.

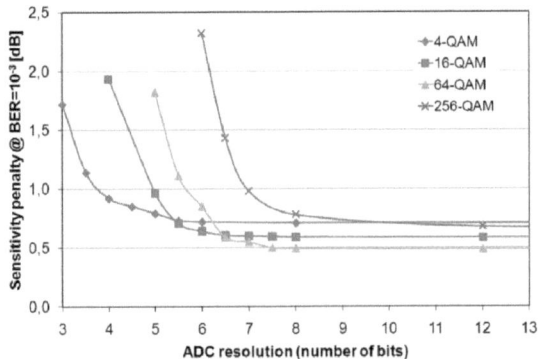

Figure 4.25: Receiver sensitivity penalty vs. analog-to-digital converter resolution
for different square QAM constellations

The necessary ADC resolution increases approximately by 1 bit if the number of constellation points is multiplied by 4. This relation becomes evident by looking again at Figure 2.3. It can be seen that increasing the number of constellations points by a factor of 4 doubles the size of the constellation diagram in real and imaginary dimensions. Therefore, to keep the accuracy of the received samples constant, the number of ADC quantization steps must also double. Table 4.3 summarizes the ADC requirements for a polarization-multiplexed 100GbE transmission system. Because commercial systems will also contain PMD and CD compensation, which necessitates oversampling, the values for $T_S/2$ sampling are also given.

4 Simulation Results

Table 4.3: Analog-to-digital converter requirements for a polarization-multiplexed QAM transmission system for 100GbE

Square constellation	ADC bandwidth	ADC sampling rate ($T_S/2$ sampling)	ADC effective number of bits
4-QAM	28 GHz	56 Gsample/s	> 3.8
16-QAM	14 GHz	28 Gsample/s	> 4.9
64-QAM	9.33 GHz	18.67 Gsample/s	> 5.7
256-QAM	7 GHz	14 Gsample/s	> 7.0

4.2.5 Square QAM internal resolutions

Not only the external quantization limited by the ADC resolution constrains algorithm performance, but also internal resolutions have to be considered. Therefore an optimal compromise must be found between performance degradation and hardware efficiency. Especially because the proposed algorithm uses B parallel blocks per module, the hardware efficiency of each block is crucial for the practicality of the system.

To find an efficient hardware implementation, one important goal is to avoid multipliers in the system, since they usually utilize a lot of chip area. Figure 4.26 shows the receiver sensitivity penalty against different resolutions of $\text{Re}[d_{k,b}]$ and $\text{Im}[d_{k,b}]$. The results are similar for all considered constellations and show that a resolution ≥ 4 bits is sufficient. As for $|d_{k,b}|^2$ the penalty for a resolution ≥ 5 bits is tolerable (Figure 4.27), the $(\ )^2$-operation in equation (3.41) can be realized with a small look-up table (4 bit input, 4 bit output) or simple logic functions.

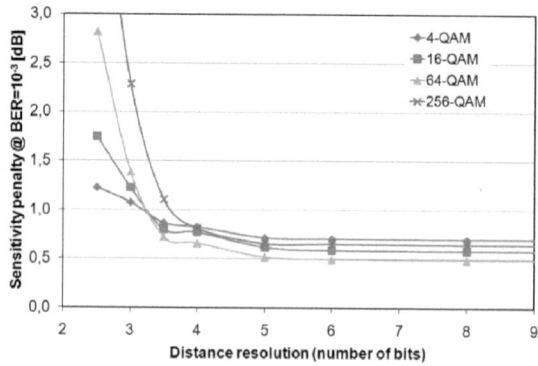

Figure 4.26: Receiver sensitivity penalty vs. internal resolution of the distances $\text{Re}[d_{k,b}]$ and $\text{Im}[d_{k,b}]$ for different square QAM constellations

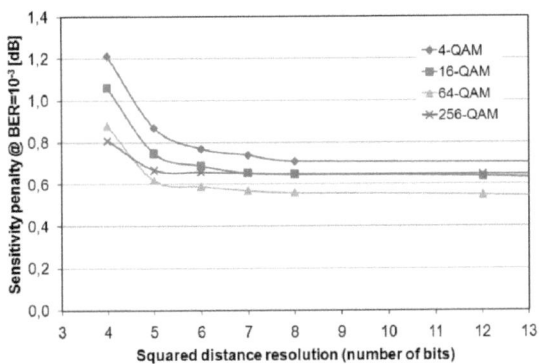

Figure 4.27: Receiver sensitivity penalty vs. internal resolution of the squared distance $|d_{k,b}|^2$ for different square QAM constellations

The reason for the similar results for all simulated constellations is that the distance to the closest constellation point is independent of the number of constellation points. This fact was already mentioned in section 3.4.4.3.2 and shows that the hardware effort to implement the proposed algorithm only increases moderately for higher-order QAM constellations. The needed internal resolutions for calculation of $|d_{k,b}|^2$ and consequently also for the subsequent filter function are always the same.

4.3 Polarization control and PMD compensation

The polarization control algorithms presented in sections 3.3.1 and 3.3.2, and even more the extension of the decision-directed algorithm to enable also the compensation of dispersive effects described in section 3.3.3 require a careful determination of the algorithm parameters to achieve an optimum performance. Additionally the simulations verify the functionality of the novel ISI compensation algorithm and demonstrate that it can improve the receiver sensitivity if the signal was corrupted by PMD.

$T_S/2$-spaced sampling is implemented. For carrier recovery the weighted Viterbi & Viterbi algorithm is chosen with a joined carrier recovery for both polarizations and $N_{CR} = 4$. The linewidth-times-symbol-duration product is set to $\Delta f \cdot T_S = 10^{-4}$.

4.3.1 Comparison of polarization control algorithms

This section compares the polarization control algorithms presented in the sections 3.3.1 and 3.3.2. The Jones matrix governing the polarization cross-talk between the two polarization modes is arbitrarily generated using a random number generator. Each data point is based on the simulation of 500,000 symbols.

4 Simulation Results

Figure 4.28 compares the influence of the control gain g on the receiver sensitivity for the non-data-aided constant modulus algorithm (CMA) and the decision-directed (DD) algorithm. The higher the control gain, the faster the polarization controller can track polarization changes.

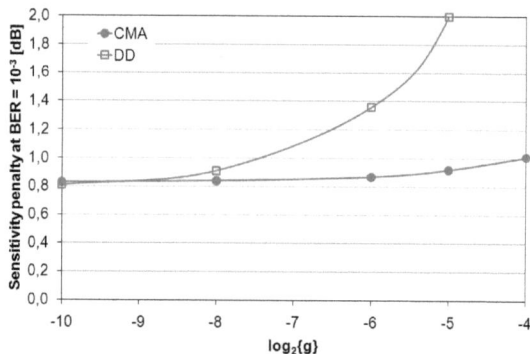

Figure 4.28: Influence of the polarization control gain g on the receiver sensitivity

The CMA tolerates a much higher control gain than the decision-directed polarization control. For the latter a 1 dB sensitivity penalty is already observed for $g \approx 2^{-7.5}$, whereas for the CMA this penalty is reached only for $g \approx 2^{-4}$. This seems to indicate that the non-data-aided algorithm allows for a $2^{3.5} \approx 11$ times faster control of the polarization.

But looking at Figure 4.29, which shows the start-up developing of the polarization control matrix elements, unveils that the same control gain g does not result in the same control time constant c_t. The decision-directed polarization controller converges about 2.3 times faster than the CMA based controller with the same control gain. The reason is that for the CMA the control update is not only weighted by the control gain, but also by the factors $1-|\underline{Y}_{x,k}|^2$ and $1-|\underline{Y}_{y,k}|^2$, respectively. This decelerates the settlement of the controller.

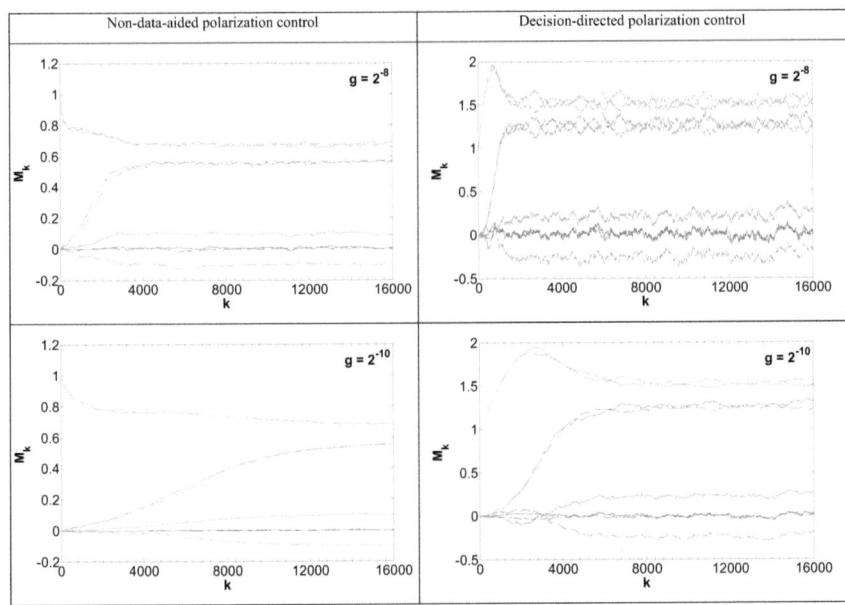

Figure 4.29: Start-up development of the matrix elements of the polarization control matrix **M**

To allow for a fair comparison of the polarization control algorithms, Figure 4.30 shows the sensitivity penalties against the normalized control time constant c_t/T_S. Now for a tolerable penalty of 1 dB the CMA allows only for a 5 times lower value of c_t. Thus the advantage of a faster polarization control as implicated from Figure 4.28 becomes less significant. A system designer has to ponder, if a faster polarization control is preferable using the CMA at the price of unaligned phases between the two polarization modes, or if the improved phase noise tolerance due to a common carrier recovery for both polarizations enabled by the decision-directed polarization control outweighs a slower polarization control.

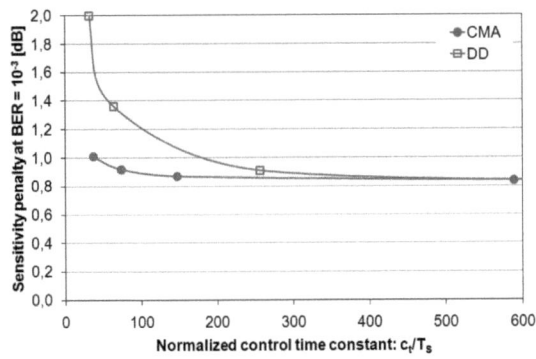

Figure 4.30: Sensitivity penalty vs. normalized polarization control time constant
for the non-data-aided and decision-directed polarization control algorithms

4 Simulation Results

4.3.2 Verification of the ISI compensation algorithm

The extension of the decision-directed polarization control algorithm to allow also the compensation of ISI caused e.g. by dispersive effects is newly presented in this book. In the following the functionality of the algorithm is verified at the example of PMD.

4.3.2.1 Impact of the control gain on the receiver sensitivity

A trade-off between control accuracy, which increases for a lower control gain g, and the control rate, that increases with increasing control gain has also to be found for the ISI compensation algorithm. Figure 4.31 shows the influence of different values of g on the receiver sensitivity.

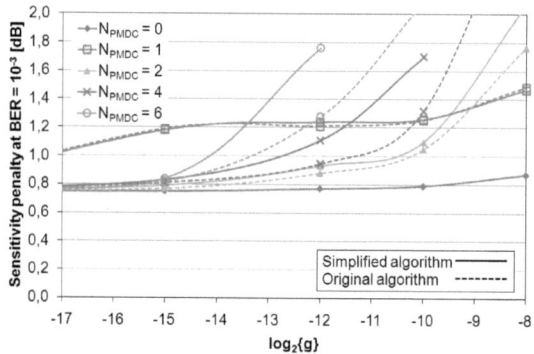

Figure 4.31: Sensitivity penalty at BER = 10^{-3} for different control gains of the polarization control/ISI compensation algorithm and different values of the FIR filter half width

The graph for $N_{PMDC} = 0$ represents the polarization control algorithm presented in section 3.3.2. It can be implemented with a relatively low control gain. For $g < 2^{-10}$ the receiver sensitivity is virtually constant, only for $g = 2^{-8}$ a small sensitivity degradation is observed.

This changes if N_{PMDC} is increased, i.e. the algorithm is extended to allow also for ISI compensation as described in section 3.3.3. The larger the filter width N_{PMDC} becomes, the higher is the required accuracy for the ISI compensation. This is apparent because $[\underline{Y}_{x,k}\ \underline{Y}_{y,k}]^T$ is calculated as the sum of $2N_{PMDC}+1$ complex control matrix multiplications. If one assumes that each control matrix is corrupted by uncorrelated AWGN with the variance σ_C^2, then additional AWGN with the variance $(2N_{PMDC}+1)\sigma_C^2$ is loaded to $[\underline{Y}_{x,k}\ \underline{Y}_{y,k}]^T$. Thus if the filter half width N_{PMDC} is increased by 1 the control gain must be reduced by $\sqrt{2}$ to compensate for the additional noise loading.

The reason that the sensitivity for $N_{PMDC} = 1$ is inferior for low control gains is that the factors $\chi_{\pm 1} = \frac{1}{2}$ do not ideally compensate for the inherent correlations between the $T_S/2$-spaced samples. Thus a residual ISI remains. For larger filters this can be compensated by the additional filter taps. For $N_{PMDC} = 1$ the residual ISI degrades the receiver sensitivity.

A comparison of the continuous lines representing the simplified control matrix update with equation (3.28) and the dashed lines representing the elaborate matrix update according to equation (3.27) shows that the latter allows for lower control gains. This is due to the fact that several correlation results are used to update one control matrix, which causes a portion of the AWGN to be averaged out. Figure 4.32 exemplifies the different receiver sensitivities for $g = 2^{-10}$.

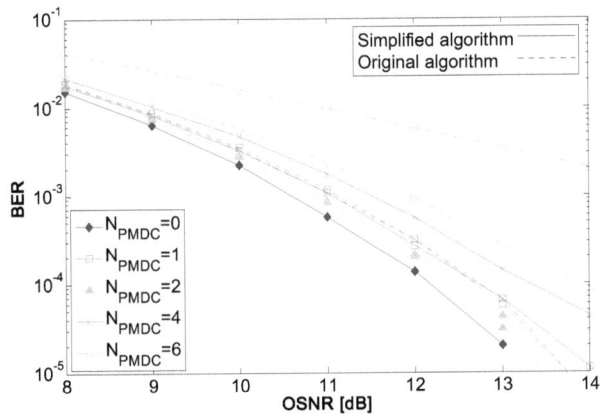

Figure 4.32: BER vs. OSNR for the simplified and original ISI compensation algorithm with $g = 2^{-10}$ and different FIR filter half widths

4.3.2.2 PMD compensation performance

To be able to evaluate the performance of the ISI compensation algorithm, a PMD emulator (PMDE) is integrated in the system simulations [29]. It implements the PMD model described in section 2.2.3.4. The MATLAB® code for the PMDE as well as the code to plot the DGD profiles were provided by Prof. Dr.-Ing. Reinhold Noé.

The transversal filter length is set to $L_{\text{PMDE}} = 10$ and $\tau_0 = \tau_1 = \ldots = \tau_{10} = T_S/4$. The average differential group delay (DGD) generated by the PMDE corresponds to $\Delta\tau_{\text{DGD}} \approx 0.8 \cdot T_S$. Figure 4.33 shows the 8 arbitrarily generated input referred DGD profiles, which are used to evaluate the PMD compensation performance of the algorithm [56].

4 Simulation Results

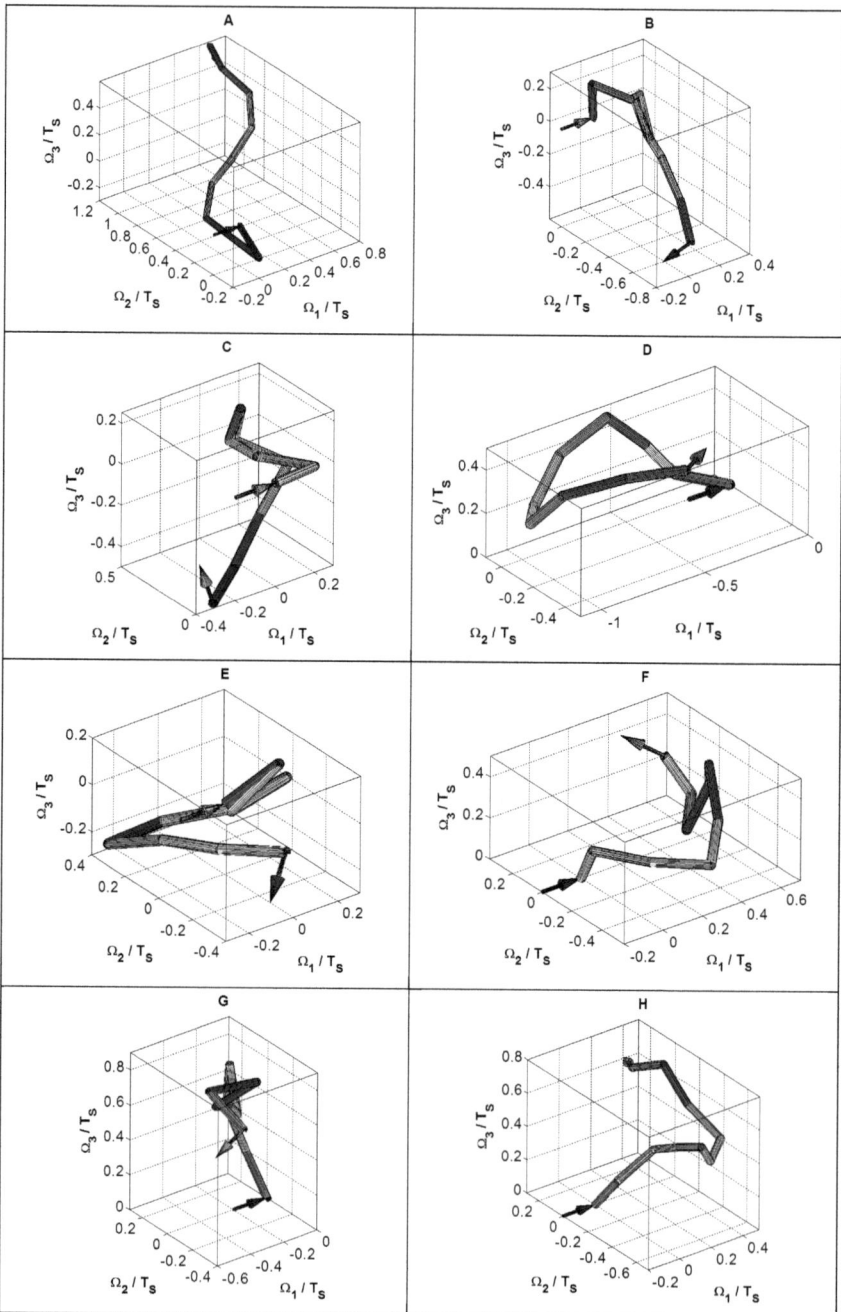

Figure 4.33: Input referred DGD profiles for the emulation of PMD

Figure 4.34 depicts the achieved receiver sensitivities after PMD compensation using the original algorithm and different FIR filter half widths. The theoretical receiver sensitivity according to equation (4.1) serves as a reference.

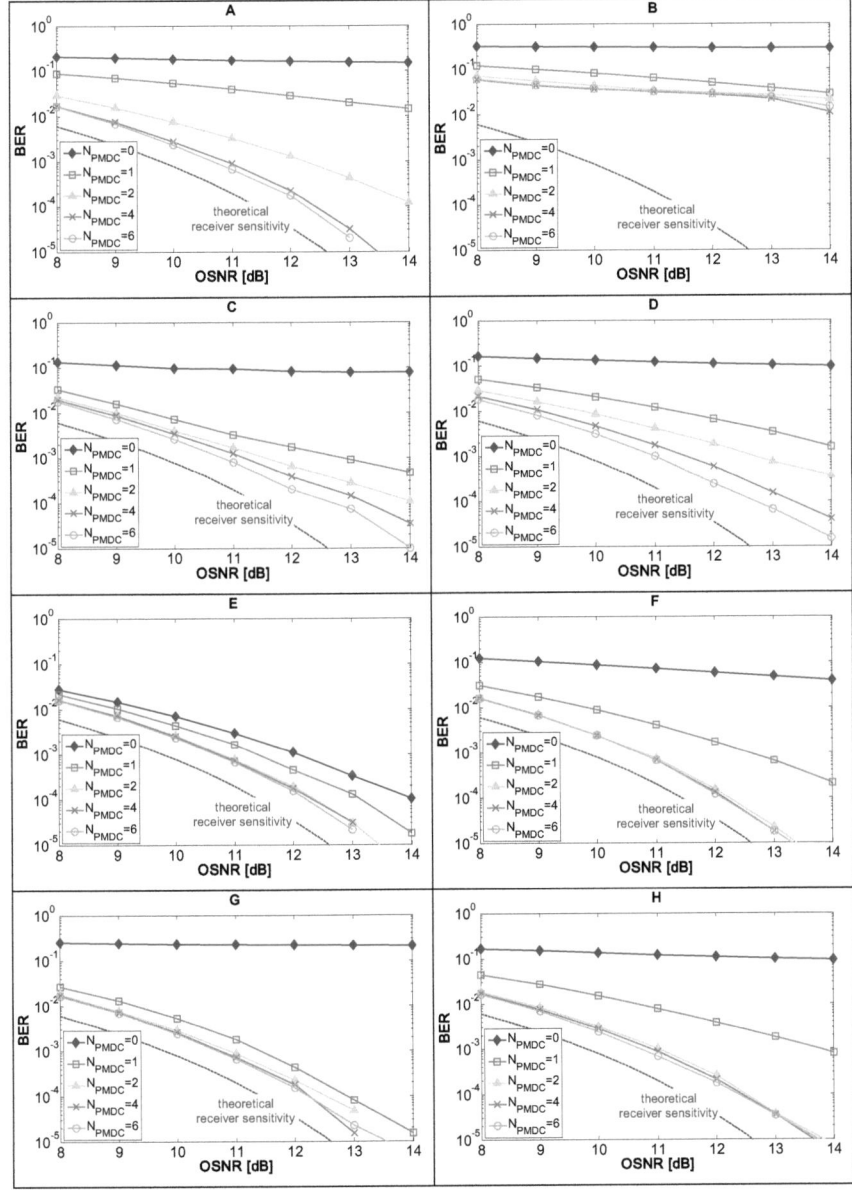

Figure 4.34: Receiver sensitivity for the original ISI compensation algorithm with different FIR filter half widths for different DGD profiles

4 Simulation Results

Figure 4.35 shows the same information Figure 4.34, but for the simplified algorithm.

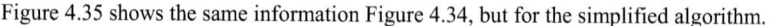

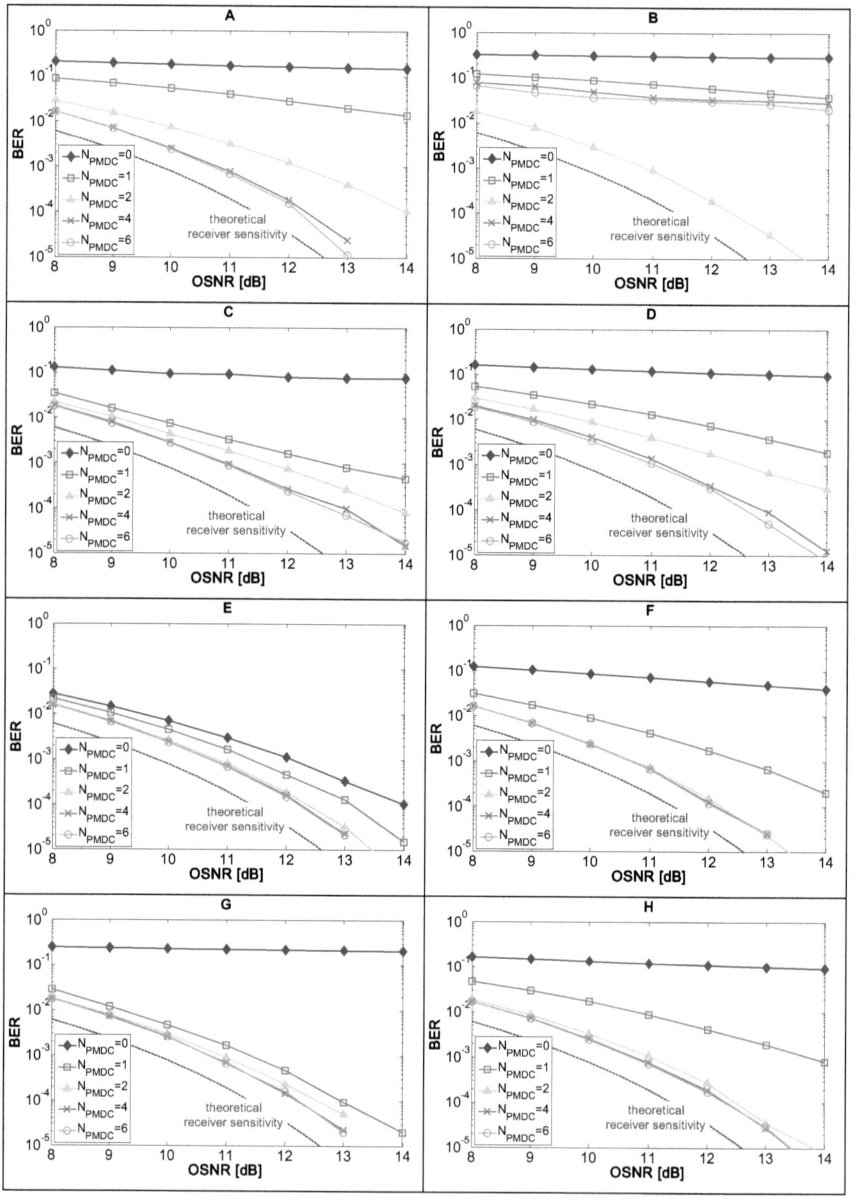

Figure 4.35: Receiver sensitivity for the simplified ISI compensation algorithm with different FIR filter half widths for different DGD profiles

The simulation results verify that both algorithms efficiently compensate ISI due to PMD. The performances of the original and simplified algorithm are thereby roughly the same. As the simplified algorithm requires a much lower computational effort, in the following only the simplified algorithm is considered.

As expected the receiver sensitivity increases with increasing FIR filter half width. For all considered DGD profiles except example B the receiver sensitivity is significantly improved and for $N_{PMDC} = 6$ almost reaches the receiver sensitivity of the PMD-free case. Figure 4.36 shows the developing of $|\det\{\mathbf{M}_i\}|$ over the $2N_{PMDC}+1$ control matrices for different values of N_{PMDC}.

4 Simulation Results

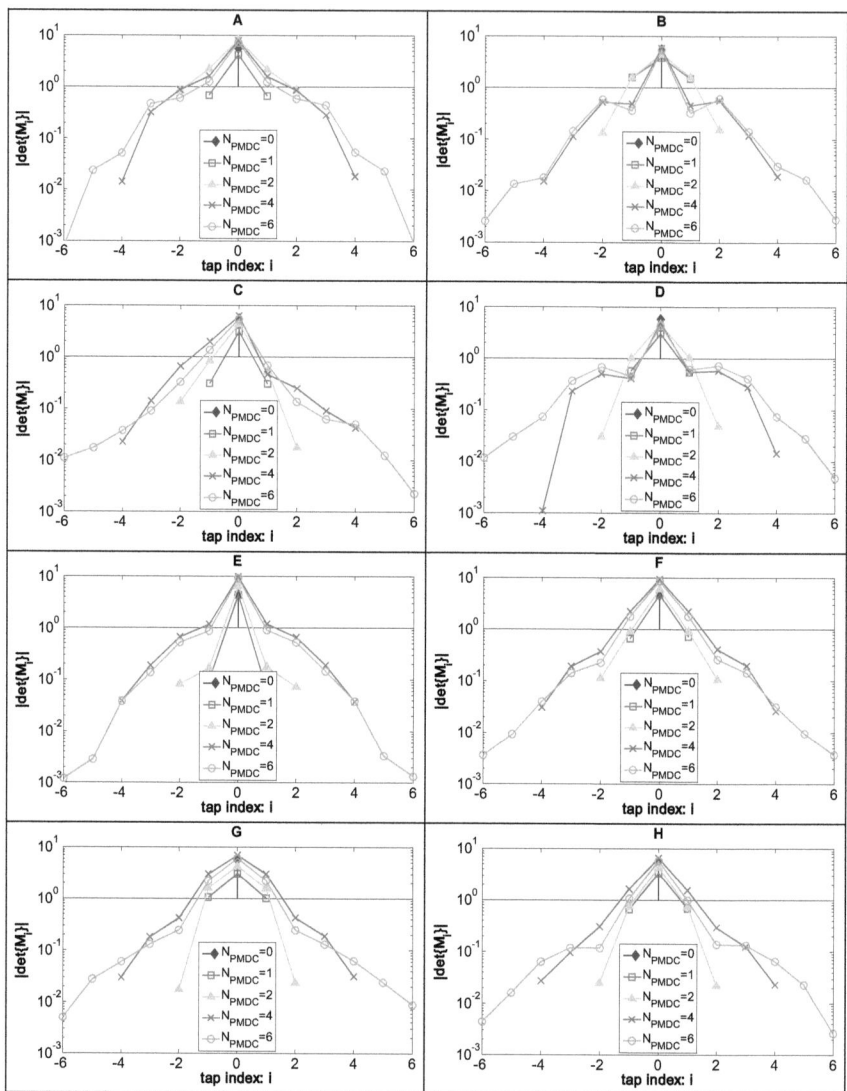

Figure 4.36: Developing of $|\det\{\mathbf{M}_l\}|$ for the simplified ISI compensation algorithm with different FIR filter half widths for different DGD profiles

The determinant of the center matrix has the highest absolute value and it decays with increasing distance to the center element. This confirms with the nature of dispersion, where the energy is spread over consecutive symbols with the same characteristic. Only the examples B and D show a deviation from this shape with $|\det\{\mathbf{M}_{\pm 2}\}| > |\det\{\mathbf{M}_{\pm 1}\}|$ for $N_{\text{PMDC}} > 2$. It is peculiar that for example B also the receiver sensitivity significantly degrades for $N_{\text{PMDC}} > 2$. Very likely this indicates that the control locks to a local minimum for the crosstalk between the different symbols, but does not find the global minimum. In order to minimize the probability of this locking to a local minimum the start-up sequence of the ISI compensator is optimized.

4.3.2.3 Optimization of the start-up sequence

Figure 4.35-B/D and Figure 4.36-B/D show that it is possible that the ISI compensation algorithm might converge to a local minimum for the crosstalk between adjacent symbols rather than to the global minimum. However, the results also indicate that for $N_{PMDC} \leq 2$ the algorithm converged to this global minimum. Thus it seems reasonable to assume that the probability that the algorithm converges to a local minimum increases with increasing N_{PMDC}. Therefore the start-up sequence of the algorithm is optimized.

The idea is that the algorithm first achieves a coarse locking to the global minimum by updating only the closest neighbor matrices of the center control matrix, and then improves the accuracy by successively activating the update for the control matrices for the more distant symbols. Figure 4.37 shows the results obtained with this optimized start-up sequence for the DGD profiles B and D.

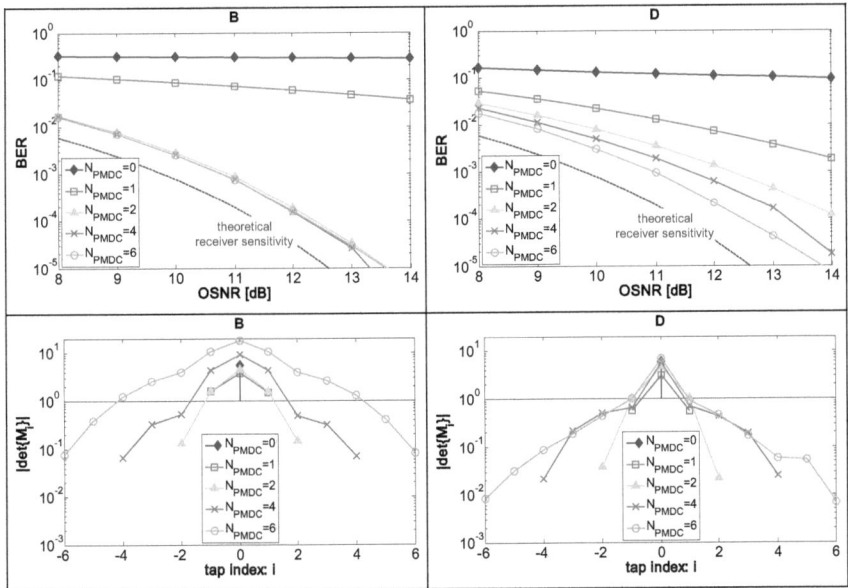

Figure 4.37: Receiver sensitivity (top row) and developing of $|\det\{\mathbf{M}_i\}|$ (bottom row) for different DGD profiles and different FIR filter half widths for the simplified algorithm with optimized start-up sequence.

The receiver sensitivities for case B with $N_{PMDC} = 4$ and $N_{PMDC} = 6$ show a significant improvement compared to the results depicted in Figure 4.34, and also the sensitivity for case B is slightly improved. The reason is that the $|\det\{\mathbf{M}_i\}|$ now continuously decay with increasing distance to the center element. Thus the algorithm converges more likely to the global crosstalk minimum.

4 Simulation Results

In the simulations the activation of the update is simply controlled by a counter, i.e. every 2^{15} symbols another matrix update is activated until all updates are active. In a practical system the effective FIR filter width could be adapted to the actual DGD profile by monitoring the determinants of the control matrices. A possible rule for increasing the filter width could be

$$\left|\det\{\mathbf{M}_l\}\right| > 2^{-5} \Rightarrow \text{ activate update for } \mathbf{M}_{l+\text{sgn}\{l\}}, \tag{4.7}$$

where l is the index of the last updated control matrix. A similar rule could be applied to reduce the filter width:

$$\left|\det\{\mathbf{M}_{l-\text{sgn}\{l\}}\}\right| < 2^{-6} \Rightarrow \text{ deactivate update for } \mathbf{M}_l \text{ and set } \mathbf{M}_l := \mathbf{0}. \tag{4.8}$$

The different switching thresholds avoid possible oscillations.

This mechanism not only solves the start-up problem, but additionally improves the receiver sensitivity. By the deactivation of redundant FIR filter taps the noise loading of the received signal by the ISI compensation filter is reduced.

5 Implementation of a synchronous optical QPSK transmission system with real-time coherent digital receiver

In the framework of the European synQPSK project a real-time synchronous QPSK transmission testbed had to be developed to verify the functionalities of the different developed components, i.e. the QPSK modulators, optical 90° hybrids, coherent receiver frontends and digital signal processing circuits, and to investigate their interplay. Thus the development of the synQPSK testbed was only possible in a broad cooperation between a multitude of researchers. The focus within this book is placed on the experimental results achieved with the testbed, as my main responsibility was the final assembly of the testbed and the execution of the measurements. Although I also supported the DSPU development, especially the compilation of the register transfer level (RTL) description of the algorithms in the very high speed integrated circuit (VHSIC) hardware description language (VHDL), these results are not addressed in this book. For detailed information about the VHDL development I refer to [57; 58; 59].

5.1 Single-polarization synchronous QPSK transmission with real-time FPGA-based coherent receiver

The first assembled testbed was mainly built from commercial components. The only applied component developed within the synQPSK project was the optical 90° hybrid provided by CeLight Israel [60]. The purpose of the testbed was to verify the VHDL code developed for the CMOS chip version A, which should contain the carrier & data recovery for a single-polarization synchronous QPSK transmission system.

5.1.1 Single-polarization synchronous QPSK transmission setup

In this subsection the assembly of the single-polarization synchronous QPSK transmission setup is described. Figure 5.1 depicts the structure of the testbed. In the following the different components are described in more detail.

5 Implementation of a synchronous optical QPSK transmission system with real-time coherent

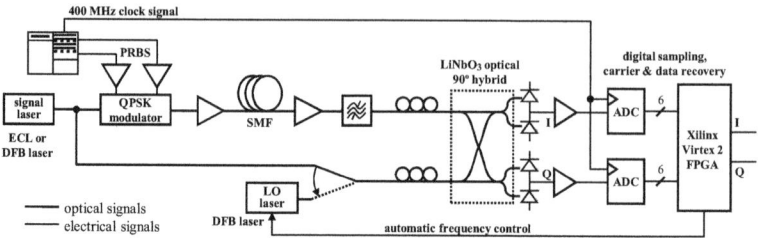

Figure 5.1: 800 Mb/s single-polarization synchronous QPSK transmission setup
with real-time digital coherent receiver

5.1.1.1 QPSK transmitter

The output signal of a 192.5 THz DFB laser (JDSU) with a specified linewidth of $\Delta f_{3dB,DFB} = 1\,\text{MHz}$ was fed into a fiber-pigtailed QPSK modulator (Bookham) to generate the QPSK optical signal. As an alternative signal laser also a tunable external-cavity laser (ECL) with a lower linewidth of $\Delta f_{3dB,ECL} = 150\,\text{kHz}$ was available. A pattern generator was used to generate a 400 Mb/s pseudo random binary sequence (PRBS) data stream with a switchable PRBS pattern length of 2^7-1 and $2^{31}-1$. The 400 Mb/s data stream was split and mutually delayed by 30 ns, which corresponds to 12 bit durations at 400 Mb/s, to emulate two decorrelated patterns. For experimental convenience the differential precoder described in section 2.2.1, that is normally required in a QPSK transmitter was omitted. Two modulator drivers (TriQuint) were used to drive the QPSK modulator with the two 400 Mb/s data streams (I and Q). Thus the modulator output was an optical 2x400 Mb/s NRZ-QPSK signal, which was then fed into the transmission fiber. Figure 5.2 shows a photograph of the QPSK transmitter.

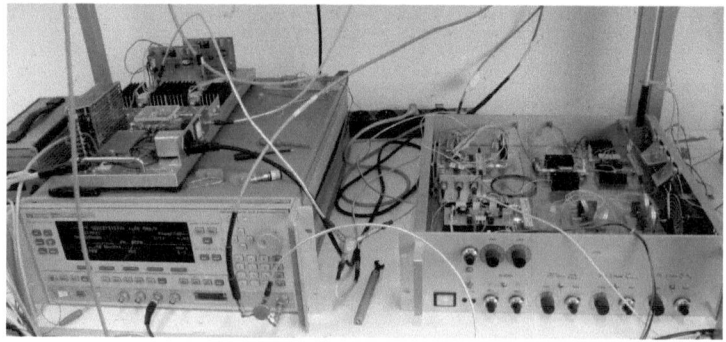

Figure 5.2: Single-polarization QPSK transmitter with Bookham optical QPSK modulator.

5.1.1.2 Coherent optical receiver frontend

After transmission through either 2 km or 63 km of standard Single-mode fiber (SMF) [ITU-T G.652] and a variable optical attenuator (VOA) to control the optical power the signal was fed

digital receiver

into the coherent optical receiver frontend. It employed an erbium-doped fiber amplifier (EDFA) as an optical preamplifier followed by a dense wavelength division multiplexing (DWDM) arrayed-waveguide grating (AWG) demultiplexer (DEMUX) with Gaussian passbands and 100 GHz channel spacing. The DEMUX output signal was fed into a second EDFA which was used for power control followed by a bandpass filter with a width of ~20 GHz. Then the received signal was fed into a LiNbO$_3$ optical 90° hybrid provided by CeLight Israel, where it was superimposed with the local oscillator (LO) signal provided from a second 192.5 THz DFB laser (JDSU) with $\Delta f_{3dB,DFB} = 1\,\text{MHz}$ for intradyne operation or from the transmitter laser for self-homodyne operation. The polarizations of the LO and received signals were matched manually by using quarter-wave plate arrays. The outputs of the optical 90° hybrid were detected in two differential photodiode pairs. Their output currents were converted to voltage signals through resistive loads and amplified in two 10 GHz bandwidth amplifiers (Picosecond).

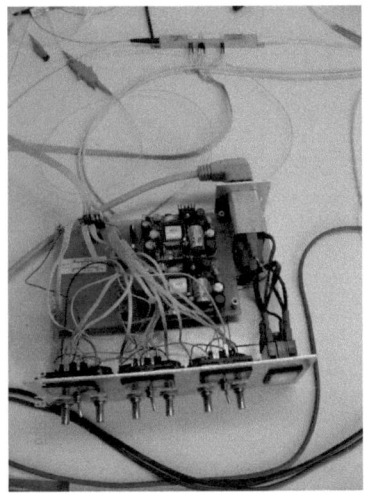
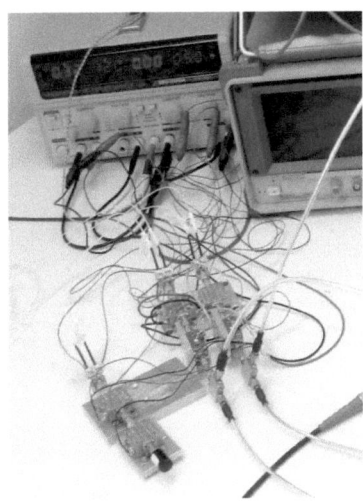

Figure 5.3: LiNbO3 optical 90° hybrid and associated control unit

Figure 5.4: Differential photodiode pairs and amplifiers

5.1.1.3 FPGA-based digital signal processing

To digitize the received signal a MAX105EVAL evaluation board (Maxim) was used. It was equipped with a MAX105 dual 6-bit analog-to-digital converter (ADC). It converted the analog signals of the I and Q component to digital outputs at up to 800 MS/s with a 400 MHz, -0.5 dB analog input bandwidth. Its -3 dB analog input bandwidth is 1.5 GHz. The dual ADC board provided LVDS digital outputs with an internal 6:12 demultiplexer that reduced the output data rate to one half the sample clock rate. This allowed easier interfacing with the subsequent digital signal processing unit (DSPU). Data was output in two's complement format. In the experiment the ADCs sampled the analog input signal at the symbol rate, i.e. with 400 MS/s.

The carrier & data recovery was implemented on a Xilinx Virtex II prototype platform populated with a Xilinx Virtex II FPGA (XC2V2000). The board offered a sufficient number of differential user I/Os to the FPGA, which were fast enough to accept the 200 Mb/s LVDS output data streams from the ADC. The FPGA further demultiplexed the input data into 16 parallel channels. This reduced the clock frequency for the FPGA core to 25 MHz. The core contained the carrier and data recovery as described in section 3.4.3 with the quantizations determined in section 4.1. Additionally an external IF control was implemented as described in section 3.6.1. The recovered data was reassembled to two full-rate serial bit streams that were analyzed in an external bit error rate tester (BERT). Figure 5.5 shows a photograph of the ADC and FPGA board.

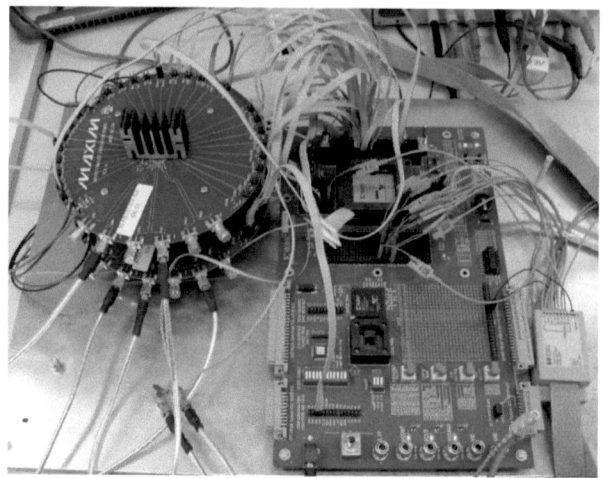

Figure 5.5: Commercial ADC board (left) and FPGA board (right) for digital signal processing

The VHDL code for the FPGA was developed in cooperation between SCT and ONT. SCT developed the core program describing the carrier & data recovery according to the specifications provided by ONT [57]. In consideration of the required hardware effort for the different algorithms (Table 3.2) and the simulation results of section 4.1 the SMLPA algorithm with $N_{CR} = 3$ was chosen for implementation. The I/O logic, clock management, IF control and monitoring functions were developed at ONT.

5.1.2 Self-homodyne experiment results at 800 Mb/s

The first test of the system was conducted with a 2^7-1 PRBS as a self-homodyne experiment at data rates of 600 Mb/s (300 Mbaud) and 800 Mb/s (400 Mbaud) using the external-cavity laser (ECL). Figure 5.6 depicts the achieved BER. I&Q channel behavior is very similar. At both data rates transmission was error-free during a 30 min test with -37 dBm of received optical power. The measured receiver sensitivity at BER = 10^{-3} is about −48 dBm for 600 Mb/s data rate and −49 dBm for 800 Mb/s data rate. Note that the applicable sum linewidth-times-symbol-duration products of 10^{-4} and $7.5 \cdot 10^{-5}$ can be achieved with commercial DFB lasers in a 10 Gbaud system.

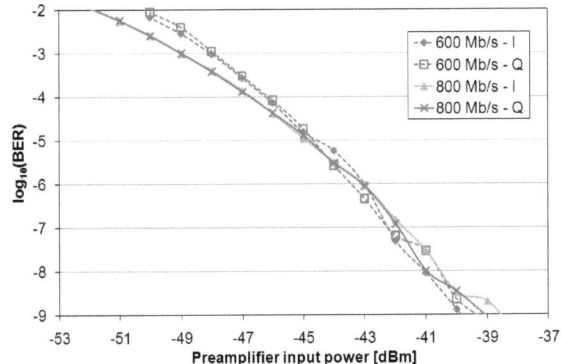

Figure 5.6: BER vs. preamplifier input power for synchronous QPSK transmission with self-homodyne detection using an ECL.

Next, the ECL was replaced by the DFB laser and the transmission experiment was repeated with data rates of 600 Mb/s, 667 Mb/s, 733 Mb/s and 800 Mb/s. Figure 5.7 shows that due to the larger linewidth of the employed laser BER floors emerge. For simplicity only averaged I&Q channel BERs are plotted. The minimum achieved BERs are $9.2 \cdot 10^{-3}$, $6.9 \cdot 10^{-3}$, $4.5 \cdot 10^{-3}$ and $3.8 \cdot 10^{-3}$, respectively.

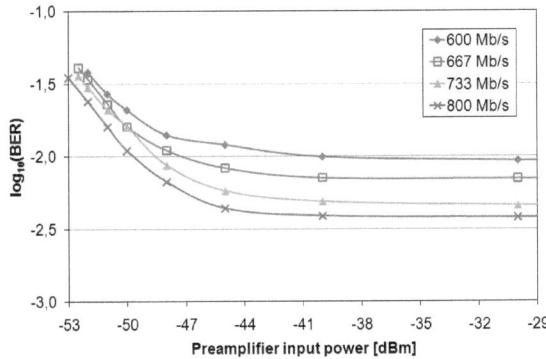

Figure 5.7: BER vs. preamplifier input power for synchronous QPSK transmission with self-homodyne detection using a DFB laser.

5.1.3 Intradyne experiment results at 800 Mb/s

Finally, an identical second DFB laser was added and used as the LO source. Without modulation, the I&Q beat signals detected at the coherent frontend outputs were recorded with a digital sampling oscilloscope in x-y-mode and are displayed in Figure 5.8.

5 Implementation of a synchronous optical QPSK transmission system with real-time coherent

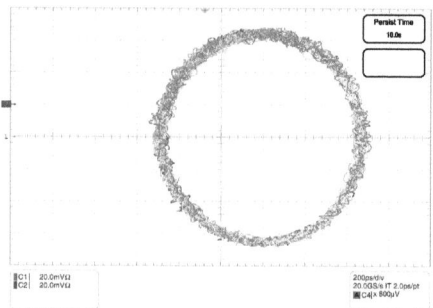

Figure 5.8: Laser beating at the output of the coherent receiver frontend

BER vs. received power is plotted in Figure 5.9 for transmission at 600 Mb/s to 800 Mb/s data rates using 2^7-1 PRBS. The best measured BER result is $6.4 \cdot 10^{-3}$ for 800 Mb/s, the highest BER floor of $1.1 \cdot 10^{-3}$ is yielded by the transmission at 600 Mb/s.

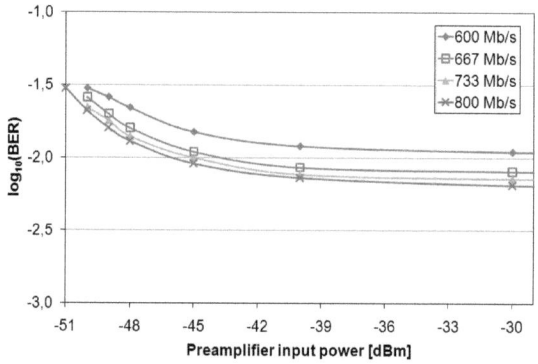

Figure 5.9: BER vs. preamplifier input power for synchronous QPSK transmission with intradyne detection using DFB lasers.

The slight increase of the BER floors compared to the self-homodyne setup is due to a broadening of the LO laser linewidth, which is most likely caused by insufficiently filtered reflections, as the laser only employs a single isolator.

5.1.4 Intradyne experiment results at 1.6 Gb/s

The bottle neck for the maximum data rate of the transmission system was the assembly of the recovered data to full-rate bit streams. As the maximum supported data rate of the FPGA I/Os was 400 Mb/s, this limited the symbol rate to 400 Mbaud. However the ADCs supported sampling rates up to 800 MS/s, and owing to the 2 times demultiplexed outputs of the ADC, the FPGA is also able to further process this data. Therefore the FPGA design was optimized and the two full-rate output bit streams were replaced by lower rate outputs that output only every fourth bit. This allowed

digital receiver

increasing the system data rate to 1.6 Gb/s, which was limited now by the maximum ADC sampling rate and the ADC-FPGA interface speed.

Figure 5.10 shows the receiver sensitivity for 1.6 Gb/s transmission over distances of 2 and 63 km using PRBS with pattern lengths of 2^7-1 (PRBS-7) and 2^{31}-1 (PRBS-31). The best measured BER was $2.7 \cdot 10^{-4}$ with PRBS-7 transmitted over 2 km, and it was $4.4 \cdot 10^{-4}$ for PRBS-31. Both PRBS could be detected until the preamplifier input power was set below -52 dBm. The BER floors for 63 km distance are slightly higher than for 2 km and are $3.4 \cdot 10^{-4}$ for PRBS-7 and $4.0 \cdot 10^{-4}$ for the PRBS-31, respectively.

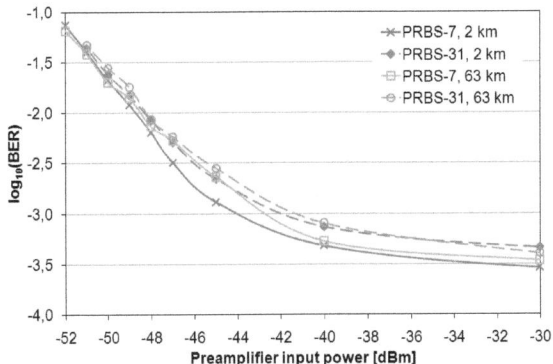

Figure 5.10: Measured BER vs. optical preamplifier input power at 1.6 Gb/s data rate.

The increase of the BER floor for longer transmission distances is most likely due to the lack of a clock recovery circuit at the receiver and the workaround of using the transmitter clock. This causes an increased clock jitter at the receiver for increasing transmission distances. The sensitivity degradation for longer PRBS patterns is most likely caused by AC coupling effects, which have a larger influence on longer patterns that contain longer '0'- or '1'-sequences.

5.1.5 System optimizations & comparison of 90° hybrid with 3x3 coupler

In order to further reduce the BER floor several system components were optimized:

- In the coherent receiver frontend the optical passband filter with a width of 20 GHz was replaced by a filter with a lower bandwidth of 15 GHz to reduce the thermal frontend noise.
- The filtering of the laser bias currents was improved to reduce the laser linewidth.
- Due to reliability problems in the ADC-FPGA interface the symbol rate was reduced to 700 Mbaud.
- The receiver was extended with a clock recovery circuit as described in section 3.2. Therefore an additional ADC was added to the system, which was clocked with an inverted clock to enable $T_S/2$ sampling of one of the receiver frontend output signals (see Figure 5.12).

5 Implementation of a synchronous optical QPSK transmission system with real-time coherent

Next to these optimizations an alternative receiver frontend was developed using a fused symmetric 3x3 coupler, i.e. the coupling ratio is 1:1:1. A 90° hybrid is in general realized with integrated optics [61], with free-space micro-optics [62], or in an all-fiber approach [63]. Its replacement by a standard component like a 3x3 coupler reduces the system cost and is thus interesting for commercial applications. In principle also the use of an asymmetric 3x3 coupler is possible, but the symmetric coupler allows for the suppression of direct detection terms [52].

The equivalent to the 90° hybrid output signal $\underline{Z}_k$ can be calculated from the symmetric 3x3 coupler output signals $z_{1,k}$, $z_{2,k}$, $z_{3,k}$ by applying the formula

$$\underline{Z}_k = \text{Re}\{\underline{Z}_k\} + j\,\text{Im}\{\underline{Z}_k\} = \left(\frac{2}{3}z_{1,k} - \frac{1}{3}(z_{2,k} + z_{3,k})\right) + \frac{j}{\sqrt{3}}(z_{2,k} - z_{3,k}). \qquad (5.1)$$

Figure 5.11 illustrates this transformation graphically.

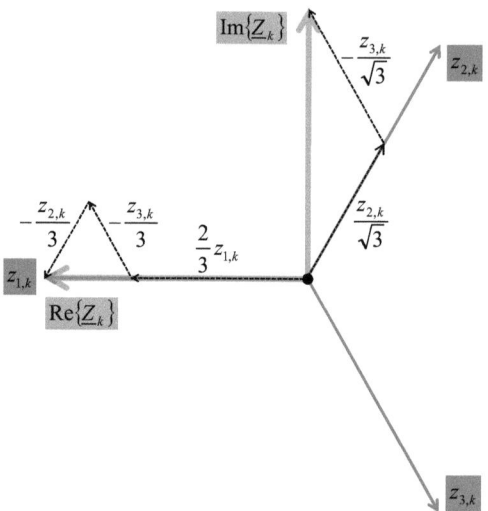

Figure 5.11: Transformation of the 3x3 coupler outputs to I&Q signals

Figure 5.12 depicts the reworked synchronous QPSK transmission setup with the 3x3 coupler employed in the coherent receiver frontend. Additionally the detailed structure for the clock recovery is depicted. The signal conversion from the 3 input signals into I&Q signals according to (5.1) is implemented inside the FPGA. For simplicity the factor $\sqrt{3}$ is approximated as 1.75.

digital receiver

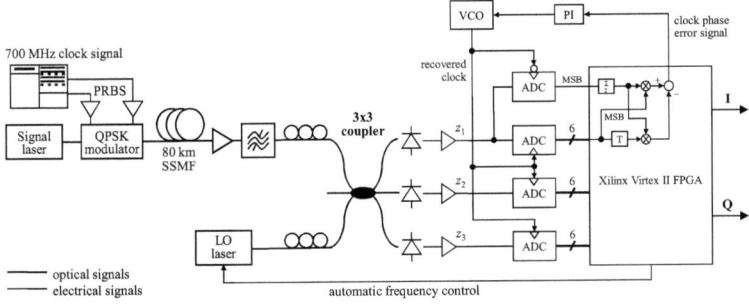

Figure 5.12: Synchronous QPSK transmission setup with clock recovery and symmetrical 3x3 coupler

The averaged BER of the received signal against the preamplifier input power at the receiver is shown in Figure 5.13. The minimum BER was $2.8 \cdot 10^{-5}$ for the receiver configuration with symmetric 3x3 coupler. For comparison the BER performance of a receiver with a 90° hybrid is about 1 dB higher than for the 3x3 coupler and fixed transformation function in the receiver. Also the achieved BER floor is with $1.7 \cdot 10^{-5}$ slightly lower. This results from the non-ideal coupling ratios in the fused 3x3 coupler. In contrast the 90° hybrid settings are optimal because both the coupling ratios and the phase shift can be controlled.

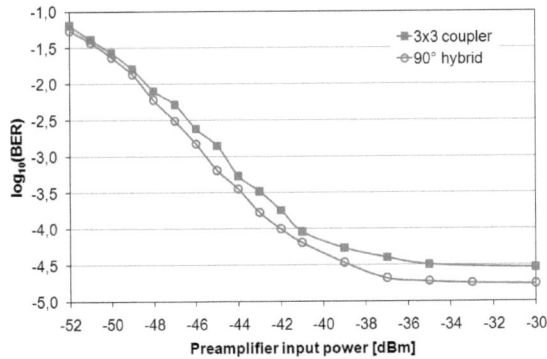

Figure 5.13: Receiver sensitivity of a synchronous QPSK receiver with either a symmetric 3x3 coupler or a 90° hybrid.

5.1.6 Comparison of experimental with simulation results

Figure 5.14 compares the measured against the simulated BER floors for the synchronous QPSK transmission system with 90° hybrid. The different measured BER floors are caused by different symbol rates that were varied from 200 Mbaud to 700 Mbaud. The corresponding linewidth-times-symbol-duration product $\Delta f \cdot T_S$ was calculated by using the specified linewidth of the DFB lasers, i.e. $\Delta f = 2 \cdot \Delta f_{3dB,DFB}$.

5 Implementation of a synchronous optical QPSK transmission system with real-time coherent

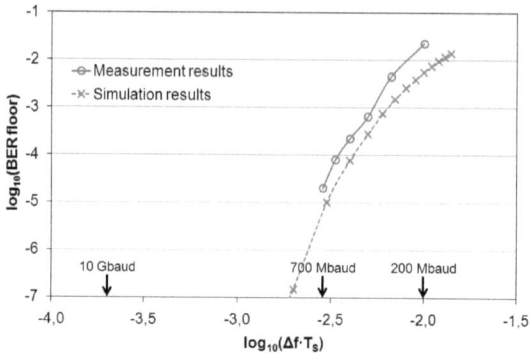

Figure 5.14: Simulated and measured BER floors for different linewidth-times-symbol-duration products

The experimental results are in good accordance with the simulations. The slight deviations can be attributed to fact that the exact linewidth was not determined. Figure 5.14 shows that at 10 Gbaud operations a BER floor due to phase noise will not be detectable any more.

5.2 Polarization-multiplexed synchronous QPSK transmission with real-time FPGA-based coherent receiver

In order to extend the testbed to polarization-multiplexed synchronous QPSK transmission a reconstruction of the transmitter and coherent receiver became necessary, because most of the components were required twice. A completely new polarization-multiplexed QPSK receiver was assembled employing two QPSK modulators provided by Photline [64]. At the receiver side a second optical 90° hybrid (CeLight) was added. Additionally to preparing the system for 10 Gbaud operation all photodiodes and amplifiers employed in the receiver were replaced by components with at least 10 GHz bandwidth. Finally also the FPGA and the associated prototyping board were replaced by a newer, larger and faster version. The main purpose of the testbed within the synQPSK project was to verify the VHDL code developed for the CMOS chip version B, which should next to the carrier & data recovery also contain a polarization control unit to enable polarization-multiplexed synchronous QPSK transmission.

5.2.1 Polarization-multiplexed QPSK transmission setup

Figure 5.15 shows a simplified block diagram for the polarization-multiplexed synchronous QPSK testbed at 2.8 Gb/s. The QPSK transmitter is extended with differential precoding circuits. This makes the programming of the BERT with particularly calculated patterns dispensable. After transmission through 80 km of standard Single-mode fiber, the received signal power is controlled by a variable optical attenuator (VOA) and then fed into a polarization scrambler. The amplified and filtered optical signal is split by a polarization beam splitter (PBS) and superimposed with the local oscillator in two 90° hybrids. Subsequent to optoelectronic and analog-to-digital conversion

digital receiver

with 1 sample/symbol, the digitized samples are fed into the FPGA, where the digital polarization control and the carrier & data recovery are implemented. Clock recovery and local oscillator frequency control are implemented like for the single-polarization experiments described in section 5.1.

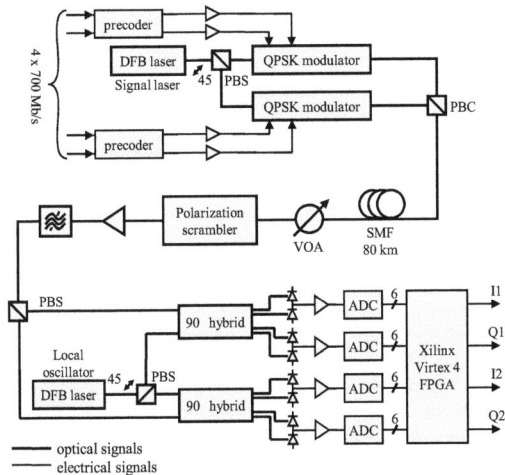

Figure 5.15: 2.8 Gb/s polarization-multiplexed QPSK transmission setup with a real-time FPGA-based synchronous coherent digital I&Q receiver

5.2.1.1 Pattern generator and QPSK precoder

Pattern generation and QPSK precoding is realized with a Xilinx RocketIO characterization board (MK325) populated with a Virtex-II ProX FPGA (XC2VP70X). The FPGA provides several multi-gigabit transceivers (MGT), that operate up to a serial data rate of 10 Gb/s per channel. The block diagram of the algorithm implemented inside the FPGA is shown in Figure 5.16.

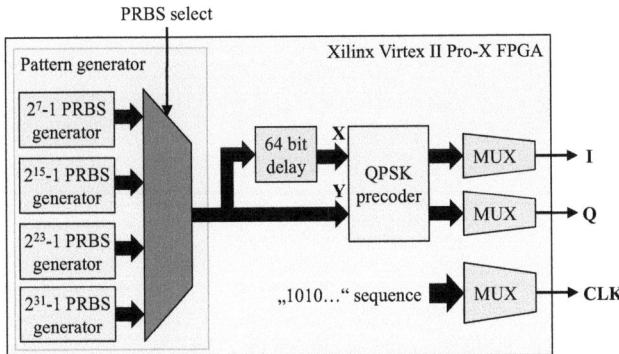

Figure 5.16: Block diagram for the pattern generator and QPSK precoder implemented in an FPGA

93

5 Implementation of a synchronous optical QPSK transmission system with real-time coherent

First the FPGA generates four different pseudo-random binary sequences (PRBS) with pattern lengths of 2^7-1, 2^{15}-1, 2^{23}-1 and 2^{31}-1 bit. By a selector controlled by two external dip switches the user can choose which pattern should be transmitted. The output of the pattern generator is split into two branches, and one branch is delayed by 64 bit for decorrelation. Then they are fed into a QPSK precoder for differential encoding according to equation (2.5). For parallel-to-serial conversion the built-in MGTs of the FPGA are used.

As the pattern generator and precoder must support symbol rates up to 10 Gbaud, parallel processing of the data has to be applied inside the FPGA. The number of parallel modules depends on the target symbol rate and ranges from 4 parallel modules for 700 Mbaud to 80 parallel modules for a symbol rate of 10 Gbaud.

Up to a symbol rate of 5 Gbaud also the transmitter clock is generated with a MGT of the FPGA by transmitting an alternating "1010…" sequence at twice the symbol rate of the system. Above 5 Gbaud only a half-rate clock can be generate with the FPGA, and an external clock multiplier has to be added to the clock output.

VHDL simulations of the program were executed with ModelSim. For VHDL synthesis, implementation and FPGA programming the Xilinx ISE^{TM} Design Suite was used. Figure 5.17 shows the MK325 Xilinx Virtex-II ProX RocketIO characterization board, on which the pattern generator and precoder are implemented.

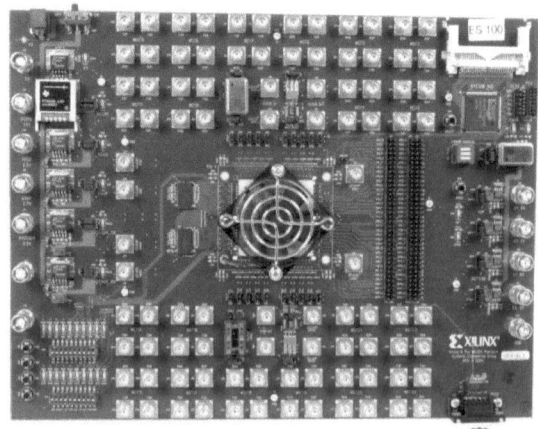

Figure 5.17: MK325 Xilinx Virtex-II ProX RocketIO characterization board
used for pattern generation and precoding

5.2.1.2 QPSK transmitter

The same 192.4 THz DFB laser (JDSU) as used for the single-polarization experiments is fed into a polarization beam splitter, whose outputs are connected to two fiber-pigtailed QPSK modulators (Photline) to generate polarization-multiplexed QPSK optical signals. Two precoded data streams are provided by the pattern generator described in section 5.2.1.1. They are fed into HF amplifiers, split and delayed to emulate decorrelated patterns for the two polarizations. The delay between the patterns for the two polarizations is generated using different cable lengths and is set to 7 symbols. Four modulator drivers (from TriQuint) are used to drive the QPSK modulators with the data streams. The QPSK modulators are followed by variable attenuators to match the powers of the two polarizations. Finally the two branches are recombined in a polarization beam combiner. The path lengths of the two branches are matched within a sub-millimeter scale. Figure 5.18 shows a photograph of the polarization-multiplexed QPSK transmitter.

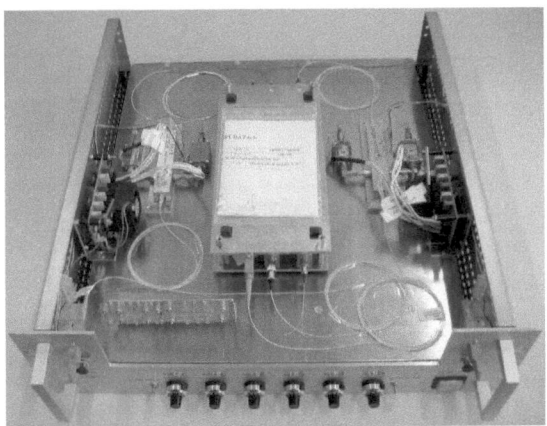

Figure 5.18: Polarization-multiplexed QPSK modulator.

5.2.1.3 Ultra-fast polarization scrambler

In order to be able to test the polarization tracking capabilities of the polarization control algorithm implemented in the coherent receiver an available polarization scrambler is integrated into the transmission channel [65]. It consists of 4 motorized fiber-optic quarter-wave plates (QWP) followed by a variable fiber-optic PDL element. Another 4 QWPs decorrelate the output SOP of the PDL element from the input SOP of a bulk-optic half-wave plate (HWP), which is followed again by 4 QWPs (Figure 5.19). The 12 QWPs rotate at different speeds ensuring that the polarization state covers all points on the Poincaré sphere and that the position dependent loss of ~2 dB of the HWP is uncorrelated with the variable PDL element. The rotation frequency of the motor that drives the HWP can be adjusted between 0 to 612 Hz. With a gearbox ratio of 39:15 from the motor to the HWP (0 to 1592 Hz) and up to 8π rad rotation on the Poincaré sphere per HWP rotation, the maximum speed of polarization changes is 40 krad/s.

5 Implementation of a synchronous optical QPSK transmission system with real-time coherent

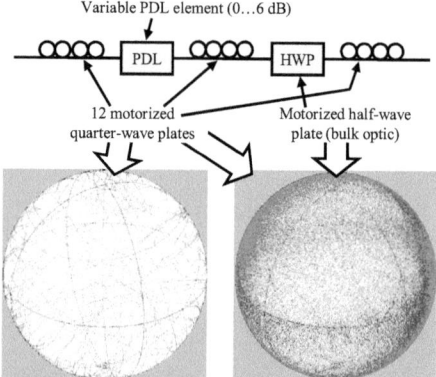

Figure 5.19: Polarization scrambler and the corresponding fast polarization changes displayed on the Poincaré sphere.

The polarization scrambler was tested by applying an unmodulated laser signal at its input and measuring the SOP at its output with a polarimeter [65]. Figure 5.19 (left) shows the SOP covering the full Poincaré sphere when only the 12 QWPs are rotating. After additionally turning on the motor driving the HWP, the Poincaré sphere fills with points (Figure 5.19 right) because the polarimeter sampling rate of 1 kHz is too slow to follow the fast polarization changes.

5.2.1.4 Polarization diversity coherent optical receiver frontend

The coherent optical receiver frontend employs an EDFA as an optical preamplifier followed by a DWDM-DEMUX with Gaussian passbands and 100 GHz channel spacing. The DEMUX output is fed into a second EDFA which is used for power control followed by a bandpass filter (BPF) with a width of ~15 GHz. Then the received signal is split by a PBS and the outputs are fed into two optical 90° hybrids, were they are superimposed with the local oscillator signal generated in a second 192.4 THz DFB laser (JDSU), which is also split by a PBS. The outputs of the 90° hybrid are detected in four photodiode pairs (Nortel). The differential signals between the photodiodes are generated by 4 differential amplifiers (Micram) with a bandwidth of 40 GHz. One output signal of the amplifier is passed through a second amplifier (Picosecond) with 10 GHz bandwidth and fed into the ADCs. The second outputs are used for monitoring the settings of the two optical 90° hybrids on two oscilloscopes. A schematic of the receiver frontend is shown in Figure 5.20.

digital receiver

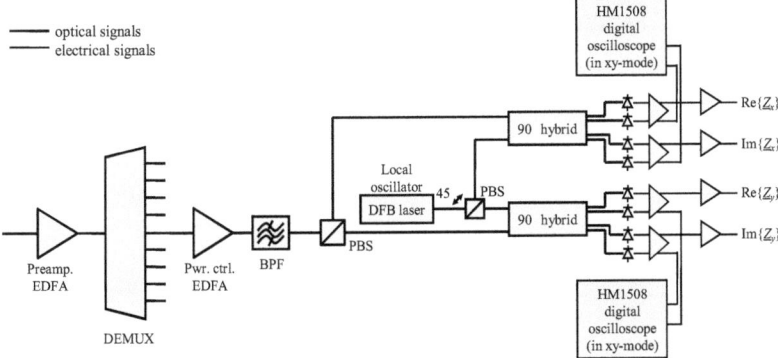

Figure 5.20: Polarization diversity coherent optical receiver frontend.

The optical path length difference between the two polarizations is matched within 1 mm. The residual delay can be matched exactly by variable delay lines connected to all four electrical outputs (I & Q of both polarizations) of the coherent receiver frontend.

5.2.1.5 FPGA-based digital signal processing

In order to evaluate the carrier & and data recovery algorithms for the CMOS chip version B, a digital coherent receiver was built up using commercially available components (Figure 5.21).

Figure 5.21: Commercially available analog-to-digital converters (left) and an FPGA board for digital signal processing (center).

To digitize the received signals two MAX105EVAL evaluation boards (Maxim) are used. The electronic polarization control and carrier & data recovery are implemented on a Xilinx Virtex 4 prototype platform equipped with a Xilinx Virtex 4 FPGA (XC4VSX35).

5 Implementation of a synchronous optical QPSK transmission system with real-time coherent

Considering the simulation results of section 4.3 a flexible implementation for the polarization control was chosen, allowing to set the polarization control gain either to $g = 2^{-4}$ or $g = 2^{-6}$. As $W = 16$ correlation results are averaged before incremental update of M, the overall system performance is similar to a system with $W = 1$ and $g = 2^{-8}$ or $g = 2^{-10}$ as simulated in section 4.3. To reduce the hardware effort only every 8th symbol is used for polarization control update. Thus according to equation (3.20) the control time constant of the system results to $c = (16/2^{-4}) \cdot (8/700 \text{ Mbaud}) \approx 3$ µs and $c = (16/2^{-6}) \cdot (8/700 \text{ Mbaud}) \approx 12$ µs, respectively. A detailed description of the VHDL program can be found in [58; 59].

Also the carrier recovery circuit is modified. The user can now select between an SMLPA filter with either $N_{CR} = 2$ or $N_{CR} = 4$.

5.2.2 Influence of different carrier recovery filter widths

At first the influence of the two different carrier recovery filters is analyzed. Figure 5.22 shows the measurement results for the two polarization channels for $N_{CR} = 2$ and $N_{CR} = 4$, i.e. 10 or 18 symbols are used for carrier recovery, respectively. The linewidth-times-symbol-duration product of the system is $\Delta f \cdot T_S \approx 2.8 \cdot 10^{-3}$.

Due to its larger bandwidth the filter with $N_{CR} = 2$ allows to achieve a lower BER floor than the filter with $N_{CR} = 4$. This result is also supported by the simulation results in section 4.1.1, which showed that for strong phase noise filters with a smaller width achieve a lower mean squared phase estimation error. As also for low preamplifier input powers the filter with $N_{CR} = 4$ does not perform better than the filter with $N_{CR} = 2$, the filter with $N_{CR} = 2$ is used for the following measurements.

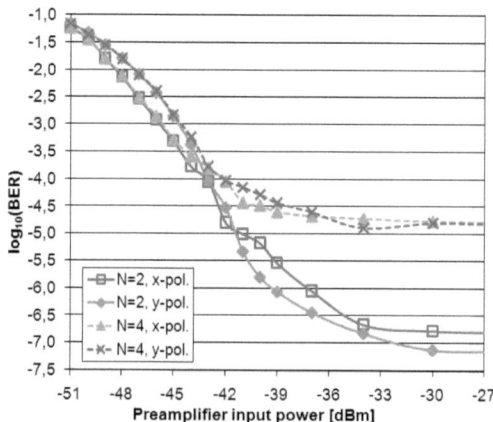

Figure 5.22: Influence of the carrier recovery filter width on the receiver sensitivity and BER floor.

A peculiarity in Figure 5.22 is that the x-polarization channel suffers from a ~1 dB penalty compared to the y-polarization channel. This effect is caused by unequal signal powers in the two polarization channels. The two QPSK modulators in the polarization-multiplexed QPSK transmitter

digital receiver

have different attenuations and the subsequent variable attenuators only allow for a coarse power matching.

5.2.3 Polarization tracking capability

Figure 5.23 shows the achieved BER against the preamplifier input power at the coherent receiver for different HWP rotation rates of the polarization scrambler described in section 5.2.1.3. The PDL of the scrambler is set to zero. For better readability the BERs of both polarizations are averaged.

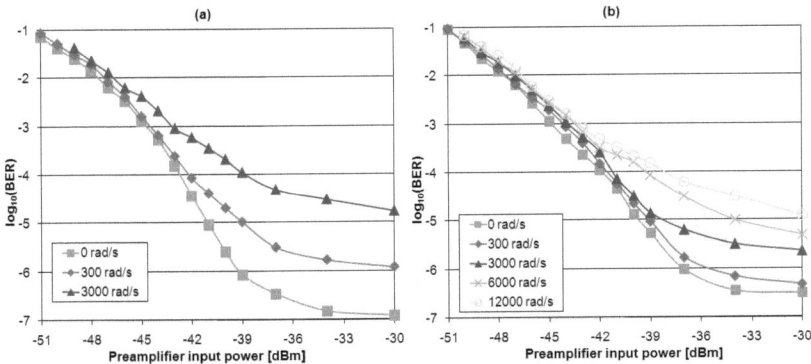

Figure 5.23: Measured BER vs. optical preamplifier input power at the coherent receiver for the two different control time constants of a) 12 µs and b) 3 µs and various polarization change speeds.

Initially the polarization scrambler is halted to measure the reference BER performance of the system. Then the scrambler is switched on and the polarization change speeds are set to 300 rad/s, 3 krad/s, 6 krad/s and 12 krad/s on the Poincaré sphere.

In Figure 5.23 (a) the polarization control time constant is set to $c = 12$ µs. The minimum bit error rate of $1.2 \cdot 10^{-7}$ was degraded by the polarization scrambler to $1.2 \cdot 10^{-6}$ and $1.7 \cdot 10^{-5}$ for the rotation speeds of 300 rad/s and 3 krad/s, respectively. With the rotation speed of 6 krad/s the BER floor was already above $1 \cdot 10^{-3}$, and for 12 krad/s the receiver was not able to compensate any more for the polarization changes.

The results for the control time constant of 3 µs are shown in Figure 5.23 (b). At $BER = 10^{-4}$ the receiver sensitivity is degraded by 0.8 dB compared to $c = 12$ µs. The minimum achievable BER of $3.1 \cdot 10^{-7}$ is also slightly worse than for $c = 12$ µs. This conforms with the simulation results of section 4.3, which also predicted a sensitivity degradation for higher control gains. However, with $c = 3$ µs the polarization scrambler degrades the BER only to $4.7 \cdot 10^{-7}$, $2.3 \cdot 10^{-6}$ and $4.8 \cdot 10^{-6}$ for the rotation speeds of 300 rad/s, 3 krad/s and 6 krad/s, respectively. Even at 12 krad/s the receiver is able to follow the polarization changes and achieves a BER of $1.2 \cdot 10^{-5}$.

Figure 5.24 shows the receiver sensitivity penalties against the polarization change speeds for $BER = 10^{-4}$. Assuming a maximum tolerable penalty of 1 dB in receiver sensitivity, the polarization

5 Implementation of a synchronous optical QPSK transmission system with real-time coherent

controller with $c = 12$ µs tolerates up to 600 rad/s. Reducing the control time constant to $c = 3$ µs increases the permissible speed of the polarization changes to 3.5 krad/s.

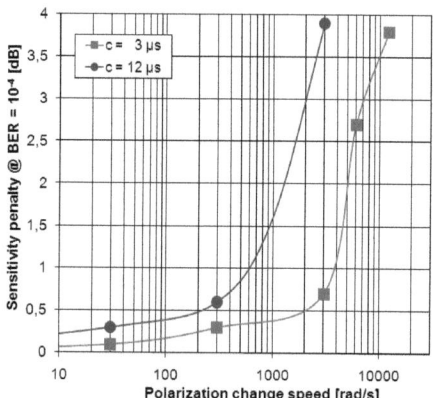

Figure 5.24: Receiver sensitivity penalty vs. scrambling speed for a bit error rate of 10^{-4}

In order to investigate if some rare events cause the polarization control to loose lock, the BER is measured for a time interval of 10 minutes with a measurement update every 10 s. The scrambling speed is chosen in such a way that it causes a 1 dB penalty in receiver sensitivity at a BER of $1 \cdot 10^{-4}$. The result is depicted in Figure 5.25. It can be seen that in spite of fast polarization change speeds of 600 rad/s and 3.5 krad/s, respectively, the BER remains almost constant. This clearly verifies the reliability of the system.

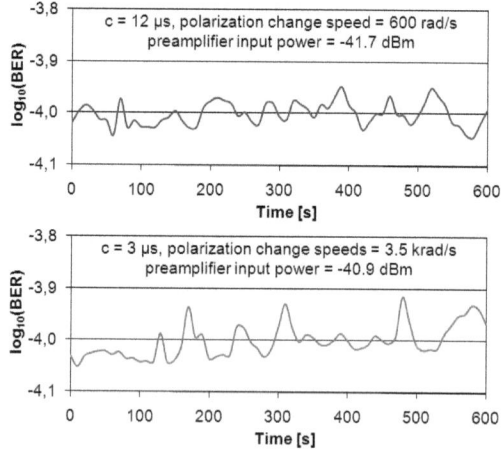

Figure 5.25: Long-term BER measurements with polarization change speeds causing 1 dB loss in receiver sensitivity.

digital receiver

5.2.4 Polarization tracking capability with optimized VHDL code

In order to achieve a faster polarization control the original VHDL code was optimized, allowing to use every second symbol in the FPGA to update the polarization control [59]. This reduces the polarization control time constant by a factor of 4, i.e. it can be switched between $c = 0.75$ µs and $c = 3.0$ µs.

First the reference BER curves had to be measured for the modified receiver. Therefore the PDL element was set to 0 dB and the BER was measured for different optical preamplifier input powers at polarization change speeds of 0, 10, 20, 30 and 40 krad/s. Figure 5.26 (a) shows the results for the accurate polarization controller with $c = 3.0$ µs, Figure 5.26 (b) depicts the corresponding values for the fast polarization control ($c = 0.75$ µs). The accurate polarization controller achieves a 0.3 dB better sensitivity at 0 krad/s than the fast one due to its higher control accuracy. Also the BER floor for $c = 3.0$ µs is with $3.8 \cdot 10^{-7}$ lower than the one for $c = 0.75$ µs, which is $6.1 \cdot 10^{-7}$. This changes dramatically if the polarization change speed is set to 40 krad/s. While the BER floor for the fast controller degrades only moderately to $4.3 \cdot 10^{-6}$ the BER floor of the accurate controller increases to $3.1 \cdot 10^{-3}$.

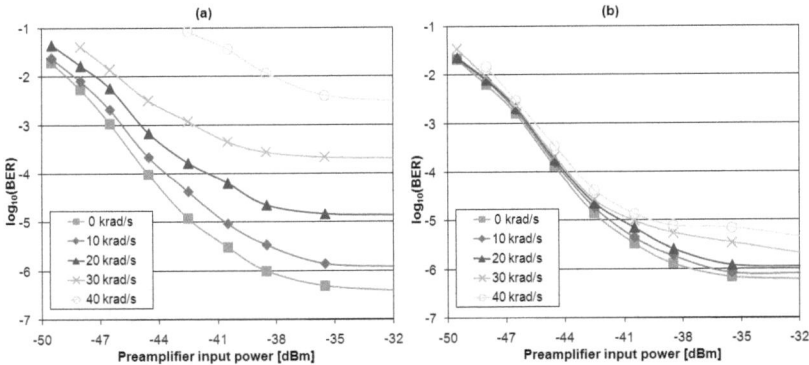

Figure 5.26: BER (I&Q averaged) vs. optical power at the preamplifier input for different speeds of polarization change. (a) for c = 3.0 µs (slow control) and (b) for c = 0.75 µs (fast control).

Figure 5.27 compares the receiver sensitivities of the two polarization controllers for different polarization change speeds at BER = 10^{-3}. As long as the polarization change speed is below ~7 krad/s the controller with $c = 3.0$ µs outperforms the one with $c = 0.75$ µs. If faster polarization changes occur the controller with $c = 0.75$ µs is advantageous. Assuming a tolerable sensitivity penalty of 1 dB the receiver with fast polarization control can tolerate polarization changes speeds faster than 40 krad/s on the Poincaré sphere, the accurate controller tolerates speeds up to 14 krad/s.

5 Implementation of a synchronous optical QPSK transmission system with real-time coherent

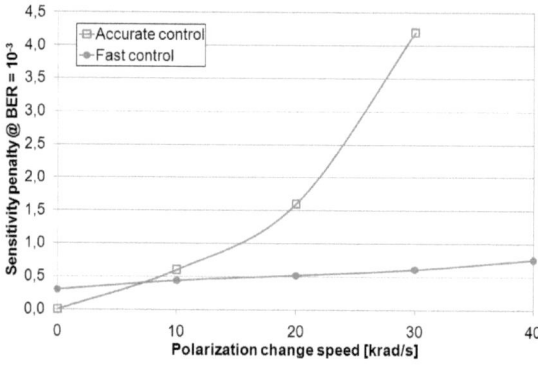

Figure 5.27: Receiver sensitivity penalty vs. speed of polarization changes on the Poincaré sphere.

5.2.5 Influence of PDL on the receiver sensitivity

In order to investigate the receiver tolerance against PDL, the PDL element of the polarization scrambler set to 6 dB. Then the BER vs. optical preamplifier input power is measured at 0 krad/s and 40 krad/s. The results are similar for both control time constants. Therefore, only the results for the fast polarization control with $c = 0.75$ µs ($g = 2^{-4}$) are shown in Figure 5.28.

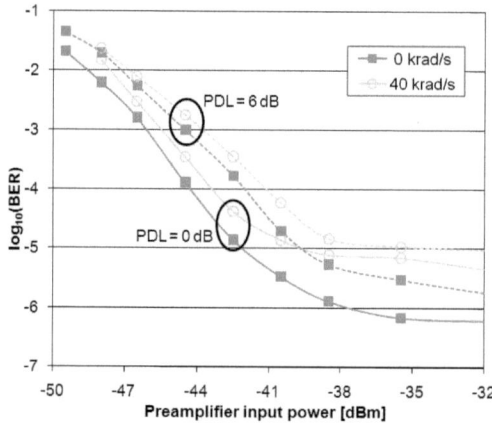

Figure 5.28: BER vs. optical preamplifier input power for 0 dB and 6 dB of PDL.

The additional penalty due to PDL is uncorrelated with the speed of the polarization changes. For 0 rad/s as well as for 40 krad/s the additional penalty due to PMD is ~1.5 dB. Figure 5.28 also shows that distortions due to PDL have a larger impact on the receiver sensitivity than polarization cross-talk, because the latter can be efficiently compensated.

5.3 Polarization-multiplexed synchronous QPSK transmission with real-time ASIC based coherent receiver

The main target of the University of Paderborn within the synQPSK project was the development of a SiGe chip for analog-to-digital conversion with a sampling rate up to 10 Gsample/s [66], and of a subsequent CMOS chip for polarization control and carrier recovery to realize a coherent synchronous polarization-multiplexed QPSK receiver that supports 40 Gb/s operations. The VHDL code for the development of the CMOS chip was verified in the experiments described in section 5.2.

The chips replace the commercial ADC boards and the FPGA, which limited the data rate of the polarization-multiplexed QPSK transmission setup depicted in Figure 5.15. Therefore four ADC chips and the CMOS chip are mounted on a common high-speed Al_2O_3 ceramic board that is placed on top of custom-designed copper blocks that act as heat sinks as well as power supply paths [67]. The whole block is placed on top of additional heat sinks. The inputs of the ADCs are connected to the coherent receiver frontend by SMA cables. Figure 5.29 shows the completed package for the SiGe and CMOS chips integrated in the testbed.

Figure 5.29: SiGe ADCs and CMOS chip package integrated in the synchronous QPSK testbed.

5 Implementation of a synchronous optical QPSK transmission system with real-time coherent

5.3.1 Transmission with and without polarization crosstalk

For the first BER measurements the data rate was set to 10 Gb/s. The polarization scrambler was halted and the polarization was set manually in such a way that the polarization crosstalk at the receiver was minimized. Then the BER was recorded for different preamplifier input powers. Afterwards the input polarization at the receiver was changed to 50% crosstalk between the two polarization channels and again the BER performance was measured. Finally the QWPs of scrambler were turned on and generated polarization changes with a speed of ~50 rad/s on the Poincarè sphere. Figure 5.30 shows the achieved bit error rates for the three cases against the input power of the preamplifier. The bit error rates of all output channels are averaged.

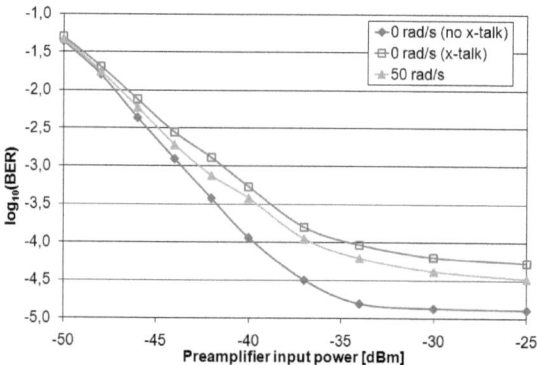

Figure 5.30: Bit error rate vs. preamplifier input power for synchronous polarization-multiplexed QPSK transmission at 10 Gb/s for different polarization states at the receiver input.

For the measurements without polarization crosstalk the best performance is achieved with a BER floor at $1.3 \cdot 10^{-5}$ and a sensitivity of -44.1 dBm for a BER of 10^{-3}. The measurements with 50% polarization crosstalk show the worst performance with a BER floor of $5.4 \cdot 10^{-5}$ and a sensitivity of -41.4 dBm for a BER of 10^{-3}. The results for polarization scrambling with 50 rad/s lie in the middle with a BER floor of $3.3 \cdot 10^{-5}$ and a sensitivity of -42.6 dBm.

The measurements show that the performance of the system is limited by the switching noise, resulting from an unsufficient filtering of power supply at the reveiver backend. The reasons for this conclusion are the following: If the polarization crosstalk is close to zero, the secondary diagonal elements of the polarization control matrix are close to zero, too. Thus the switching noise in the CMOS chip is low (best case). In contrast if the polarization crosstalk is close to 50%, all elements of the polarization control matrix toggle and the switching noise is maximal (worst case). This correlates exactly with the BER measurement results in Figure 5.30 and the observed power consumption of the CMOS chip, which increases if there is crosstalk between the polarizations. The system performance with activated scrambler consequently lies in between the best case and worst case scenario results, as due to the scrambling of the SOP the polarization cross-talk at the receiver varies between 0% and 50%.

digital receiver

5.3.2 Influence of different carrier recovery filter widths

Due to the increased data rate that is possible with the SiGe ADCs and the CMOS chip the linewidth-times-symbol-duration product changed from $\Delta f \cdot T_S \approx 2.8 \cdot 10^{-3}$ for the measurements results presented in section 5.2.2 to $\Delta f \cdot T_S \approx 8 \cdot 10^{-4}$ for the current setup. Figure 5.31 shows the influence of the filter half width N_{CR} on the receiver sensitivity.

The larger filter with $N_{CR} = 4$ yields a better system performance with a BER floor of $3.3 \cdot 10^{-5}$ and a sensitivity of -43.5 dBm for a BER of 10^{-3}. The receiver with $N_{CR} = 2$ achieves a BER floor of $9.3 \cdot 10^{-5}$ and a sensitivity of -42.0 dBm. Due to the lower linewidth-times-symbol-duration product compared to the previous measurements the larger filter with $N_{CR} = 4$ now outperforms the smaller filter with $N_{CR} = 2$ by 1.5 dB in terms of receiver sensitivity.

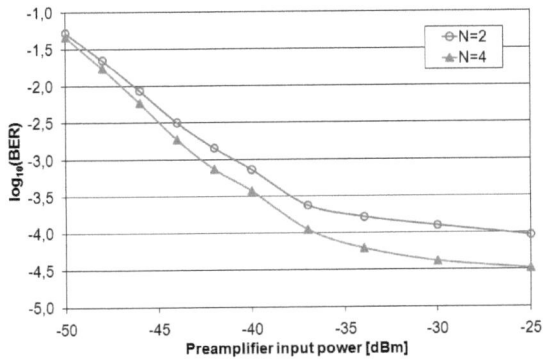

Figure 5.31: Influence of the filter half width N_{CR} in the carrier recovery circuit on the BER performance of a 10 Gb/s polarization-multiplexed synchronous QPSK transmission system

But in general a lower BER floor would be expected for $N_{CR} = 2$, because a lower filter width increases the phase noise tolerance. That this cannot be observed in the measurements shown in Figure 5.31 is due to the fact that the BER floor is not caused by phase noise but by switching noise of the CMOS chip.

5.3.3 Single-polarization vs. polarization-multiplexed QPSK transmission

In a final measurement the 10 Gb/s polarization-multiplexed QPSK transmission system is compared against a 5 Gb/s single-polarization transmission system. Thus both systems have a symbol rate of 2.5 Gbaud. Figure 5.32 shows that the polarization-multiplexed system suffers from a ~3 dB penalty compared to the single-polarization system. This is in accordance with theory because the polarization-multiplexed system sends twice the number of bits per symbol. Thus for the same preamplifier input power the energy per bit is halved compared to the single-polarization system.

5 Implementation of a synchronous optical QPSK transmission system with real-time coherent

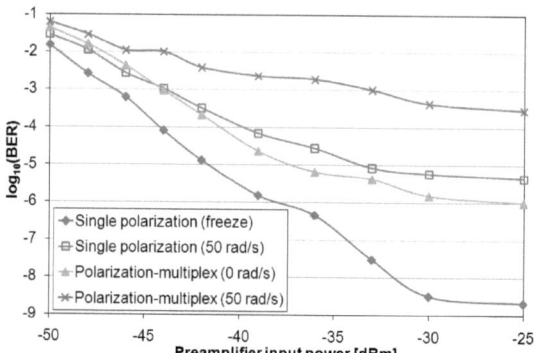

Figure 5.32: Comparison of 5 Gb/s single-polarization QPSK transmission against 10 Gb/s polarization-multiplexed QPSK transmission

The BER floors in Figure 5.32 strongly differ for the different evaluated scenarios. These large variations are caused by the switching noise of the CMOS chip, which also differs strongly for the different scenarios. For single-polarization transmission with manually optimized SOP at the receiver input the digital polarization control matrix can be frozen, i.e. the controller is disabled. This also minimizes the switching noise and a BER floor of $2 \cdot 10^{-9}$ can be achieved. As soon as the SOP at the receiver input varies due to scrambling and the digital polarization controller is switched on, the BER floor degrades to $4 \cdot 10^{-6}$. Roughly the same BER floor ($1.5 \cdot 10^{-6}$) is also achieved for polarization-multiplexed transmission with manually optimized SOP. In both cases the switching noise inside the CMOS chip is roughly the same because in both cases half of the polarization control matrix elements is zero: In the case of single-polarization transmission the elements for the second polarization are zero, for polarization-multiplexed transmission only the phase offset between the two polarizations has to be controlled and thus the secondary diagonal elements are zero. Finally if the input SOP of the polarization-multiplexed receiver is scrambled and all elements of the polarization control matrix toggle the BER floor further degrades to $3 \cdot 10^{-5}$.

6 Discussion

Until today (January 2009) most of the optical transmission experiments with coherent detection still use offline signal processing [68; 69; 70]. Although these experiments have a high value, especially for the evaluation of transmission impairments at high data rates >100 Gb/s [71], special care must be taken to ensure the feasibility of the employed algorithms for real-time implementations. The algorithmic constraints summarized in this book can help researchers to find out whether their signal processing algorithms can operate in real-time systems [72].

The most convincing way to demonstrate the real-time feasibility of an algorithm is its actual implementation in a real-time coherent receiver. The experimental results presented in this book thereby mark milestones in the development of real-time synchronous QPSK transmission systems as shown in Figure 6.1.

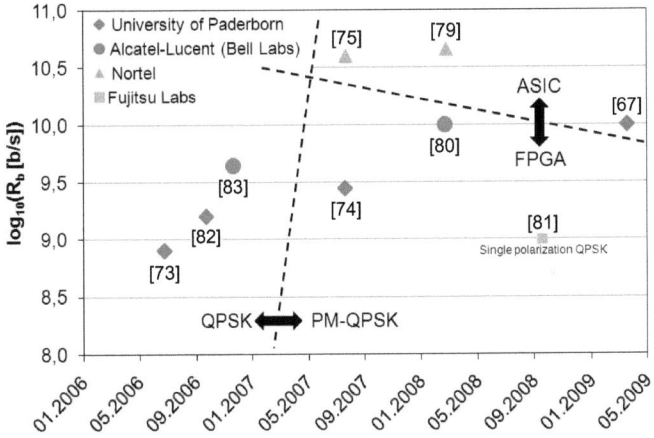

Figure 6.1: Milestones in synchronous QPSK transmission with real-time coherent digital receivers

Although researchers all over the world investigate coherent technologies for optical communication, until today only 4 research groups published experimental results obtained with real-time digital receivers. Among these the University of Paderborn was the first to achieve a single-polarization QPSK transmission with digital feed-forward carrier recovery [73] as well as a polarization-multiplexed QPSK (PM-QPSK) transmission with additional real-time digital polarization tracking [74]. As described in the sections 5.1 and 5.2 both systems were realized with commercially available ADCs and FPGAs for digital signal processing. In July 2007 the Nortel group revealed the first real-time coherent PM-QPSK receiver based on an ASIC [75]. Due to the application-specific design of the DSPU they were the first to reach the standard data rate of 40 Gb/s. The experimental results presented in section 5.3 using the ASIC-based receiver developed

6 Discussion

within the synQPSK project will be published at the Optical Fiber Communication Conference and Exposition (OFC) 2009 [67].

But the presentation of a commercial polarization-multiplexed synchronous QPSK system does not mark the end of research about coherent optical techniques. In contrast it motivates to look further. Orthogonal frequency division multiplexing (OFDM) and higher-level QAM formats promise spectral efficiencies far beyond the one of polarization-multiplexed QPSK. Integrated modulators have already been developed for 16-QAM [17; 55], and the transmission of a polarization-multiplexed 128-QAM signal was demonstrated using a pilot carrier for optical phase noise cancelling [18]. But a phase noise tolerant carrier recovery algorithm has not been proposed yet. All published algorithms only achieve a low laser linewidth tolerance in practical systems because of a decision-directed feedback loop [38; 76; 77], or the use of dedicated symbols for carrier recovery [78]. The feed-forward carrier recovery presented in this book is the first one that promises a sufficient phase noise tolerance to allow the application of DFB lasers.

7 Summary

This book has presented the different functional blocks and the corresponding algorithms that are required in a digital coherent receiver for optical transmission systems: Clock recovery, polarization control and ISI compensation, carrier recovery, data recovery and intermediate frequency control.

One emphasis was put on the comparison of two polarization control algorithms. The non-data-aided (NDA) approach uses the correlation of data before and after the polarization controller to force the output data of the controller on the unity circle. This inherently separates the two polarizations. The decision-directed (DD) approach correlates the controller output data with the recovered symbols and forces this correlation matrix to the unity matrix. Comparative simulations showed that the NDA approach allows an approximately 3 times faster tracking of the SOP, but at the price of arbitrary phase differences between the two polarization modes. In contrast the DD approach requires a larger control time constant, but also compensates for phase offsets between the polarization channels, thus allowing a common carrier recovery for both polarizations which increases the phase noise tolerance.

The DD approach for polarization control has additionally been extended to enable also the compensation of intersymbol interference (ISI). In simulations the capability of the algorithm has been successfully demonstrated at the example if PMD compensation. If the start-up sequence of the ISI compensator is correctly controlled, the algorithm successfully locked to all investigated DGD profiles and efficiently compensated for ISI. The only drawback is that for an increasing filter width the control gain has to be reduced to avoid a degradation of the receiver sensitivity. This limits the achievable tracking speed for time-variant impairments.

The comparison of different feed-forward carrier recovery schemes was another main topic within this book. Four different algorithms were investigated: The Viterbi & Viterbi (V&V) algorithm, the V&V algorithm extended by adaptive weighting of the filter inputs and the barycenter or (S)MLPA algorithm are developed for equidistant-phase constellations. A novel feed-forward algorithm presented in this book allows also for efficient carrier recovery for arbitrary QAM constellations. All algorithms were compared in QPSK simulations. An evaluation of the estimator efficiency in a phase noise-free scenario and of the mean squared estimator error in simulations considering phase noise showed that the V&V algorithm with adaptive weighting yields the best performance. However, the SMLPA algorithm is only slightly inferior, but allows for a much more efficient hardware implementation. Although the QAM carrier recovery performance is similar to the other presented approaches for QPSK, its main advantage is the efficient carrier recovery for higher-order QAM constellations. The simulation results from section 4.2 showed that with the proposed algorithm a synchronous optical 16-QAM transmission system employing DFB lasers is already feasible today, and 64-QAM or 256-QAM system will be realizable in the near future.

Finally the experimental results of a real-time synchronous QPSK transmission system developed within the European synQPSK project were presented. The simulation results presented in section

7 Summary

4.1 were thereby the main basis for the specifications of the digital signal processing unit in the real-time coherent receiver. The experiments presented in the sections 5.1 and 5.2 were the worldwide first demonstrations of a synchronous QPSK transmission system with real-time coherent digital receiver using feed-forward carrier recovery and of a synchronous polarization-multiplexed QPSK transmission system with real-time coherent receiver and digital polarization tracking, respectively. After replacing the commercial ADCs and the FPGA for signal processing in the receiver by specifically developed high-speed ADCs and a CMOS DSPU, both also developed in the context of the synQPSK project, the data rate could be increased to 10 Gb/s.

8 Outlook

The algorithms and experimental results presented in this book offer a broad variety of follow-up research:

The dispersion compensation algorithm proposed in section 3.3.3 is until now only evaluated for its PMD compensation performance. Additional simulations considering CD as well as a comparison against other state-of-the-art dispersion compensation algorithms are required to finally assess the feasibility of the algorithm for commercial systems.

The proposed QAM carrier recovery algorithm is evaluated in great detail in this book for 4 different square QAM constellations. However, many more constellations are possible and might outperform the considered constellations in terms of receiver sensitivity, transmitter and receiver complexity or tolerance against various distortions. The research about high-level QAM formats for optical communication systems is only in the early stages. Only little is known today about the tolerances against linear and non-linear impairments such as PMD and CD or self-phase modulation (SPM) and cross-phase modulation (XPM). Also the interoperability of different QAM constellations and also of the proposed feed-forward carrier recovery with other algorithms for polarization control or dispersion compensation needs to be assessed.

Finally based on the existing testbed for the evaluation of real-time synchronous QPSK transmission all algorithms described in this book, which are not yet implemented, can be translated to a hardware description language (e.g. VHDL) to be evaluated in real-time transmission experiments.

9 Bibliography

[1] **Imai, T., et al.** Field Demonstration of 2.5 Gbit/s Coherent Optical Transmission Through Installed Submarine Fiber Cables. *Electronics Letters.* August 16, 1990, Vol. 26, 17, pp. 1407-1409.

[2] **Takachio, N., Norimatsu, S. and Iwashita, K.** Optical PSK Synchronous Heterodyne Detection Transmission Experiment using Fiber Chromatic Dispersion Equalization. *IEEE Photonics Technology Letters.* March 1992, Vol. 4, 3, pp. 278-280.

[3] **Norimatsu, S., Iwashita, K. and Sato, K.** PSK Optical Homodyne Detection Using External Cavity Laser Diodes in Costas Loop. *IEEE Photonics Technology Letters.* May 1990, Vol. 2, 5, pp. 374-376.

[4] **Derr, F.** Coherent optical QPSK intradyne system: Concept and digital receiver realization. *IEEE Journal of Lightwave Technology.* 1992, Vol. 10, 9, pp. 1290-1296.

[5] **Desurvire, E., Simpson, J. and Becker, P.** High-Gain Erbium-Doped Traveling-Wave Fiber Amplifier. *Optics Letters.* 1987, Vol. 12, 11, pp. 888-890.

[6] **Mears, R., et al.** Low-Noise Erbium-Doped Fibre Amplifier at 1.54pm. *Electronics Letters.* 1987, Vol. 23, 19, pp. 1026-1028.

[7] **Kazovsky, L.** Performance analysis and laser linewidth requirements for optical PSK heterodyne communications systems. *IEEE Journal of Lightwave Technology.* April 1986, Vol. 4, 4, pp. 415-425.

[8] **Noé, R., et al.** Direct modulation 565 Mb/s PSK experiment with solitary SL-QW-DFB lasers and novel suppression of the phase transition periods in the carrier recovery. *Proc. ECOC'92.* September 27, 1992, pp. 867-870.

[9] **Ip, E., et al.** Coherent detection in optical fiber systems. *Optics Express.* January 2008, Vol. 16, 2, pp. 753-791.

[10] **Noé, R.** Phase noise tolerant synchronous QPSK receiver concept with digital I&Q baseband processing. *Proc. OECC/COIN'04.* July 12-16, 2004, 16C2-5.

[11] **Taylor, M.** Coherent detection method using DSP to demodulate signal and for subsequent equalisation of propagation impairments. *Proc. ECOC'03.* September 21-25, 2003, We4.P.111.

[12] **Noé, R.** PLL-Free Synchronous QPSK Polarization Multiplex/Diversity Receiver Concept with Digital I&Q Baseband Processing. *IEEE Photonics Technology Letters.* April 2005, Vol. 17, 4, pp. 887-889.

[13] **Savory, S.** Compensation of Fibre Impairments in Digital Coherent Systems. *Proc. ECOC'08.* 24-28. September 2008, Mo.3.D.1.

[14] *Annex I - "Description of work" to contract no. FP6-004631.* project synQPSK, Univ. Paderborn. March 31, 2004.

[15] **Kahn, J. and Ho, K.** Modulation and Detection Techniques for DWDW Systems. [book auth.] E. Forestieri. *Optical Communication Theory and Techniques.* s.l. : Springer US, 2005, pp. 13-20.

[16] **Ly-Gagnon, D., Katoh, K. and Kikuchi, K.** Unrepeatered optical transmission of 20 Gbit/s quadrature phase-shift keying signals over 210km using homodyne phase-diversity receiver and digital signal processing. *Electronics Letters.* February 17, 2005, Vol. 41, 4, pp. 59-60.

[17] **Sakamoto, T., Chiba, A. and Kawanishi, T.** 50-Gb/s 16 QAM by a quad-parallel Mach-Zehnder modulator. *Proc. ECOC'07.* September 16-20, 2007, PD 2-8.

[18] **Goto, H., et al.** Polarization-multiplexed 1 Gsymbol/s, 128 QAM (14 Gbit/s) coherent optical transmission over 160 km using a 1.4 GHz Nyquist filter. *Proc. OFC/NFOEC'08.* February 24-28, 2008, JThA45.

[19] **Proakis, J.** *Digital Communication.* 4. s.l. : McGraw-Hill Higher Education, 2000.

[20] **Gray, F.** *Pulse code communication.* 2,632,058 USA, March 17, 1953.

[21] **Webb, W. and Hanzo, L.** *Modern Quadrature Amplitude Modulation.* s.l. : Pentech Press, 1994.

[22] **Weber, W.** Differential Encoding for Multiple Amplitude and Phase Shift Keying Systems. *IEEE Transactions on Communications.* March 1978, Vol. 26, 3, pp. 385-391.

[23] **Cacciamani, E. und Wolejsza, C.** Phase-Ambiguity Resolution in a Four-Phase PSK Communications System. *IEEE Transactions on Communication Technology.* December 1971, Bd. 19, 6, S. 1200-1210.

[24] **Agrawal, G. P.** *Fiber-Optic Communication Systems.* s.l. : John Wiley & Sons, Inc., 1992.

[25] **Jones, R. C.** New calculus for the treatment of optical systems. *Journal of the Optical Society of America.* 1941, Vol. 31, pp. 488-493.

[26] **Krummrich, P., et al.** Field trial results on statistics of fast polarization changes in long haul WDM transmission systems. *Proc. OFC/NFOEC'05.* March 6-11, 2005, OThT6.

[27] **Pfau, T.** *Dispersion Compensation for Optical Phase Modulation Schemes with RZ and NRZ Impulse Shaping.* Stuttgart : University of Stuttgart, 2004.

[28] **Alwayn, V.** *Optical Network Design and Implementation.* s.l. : Cisco Press, 2004.

[29] **Noé, R., et al.** Polarization mode dispersion compensation at 10, 20 and 40 Gb/s with various optical equalizers. *IEEE Journal of Lightwave Technology.* September 1999, Vol. 17, 9, pp. 1602-1616.

[30] **Saleh, B. and Teich, M.** *Fundamentals of Photonics.* New York : John Wiley & Sons, 1991.

[31] **Calabrò, T., et al.** An electrical polarization-state controller and demultiplexer for polarization multiplexed optical signals. *Proc. ECOC'03.* September 21-25, 2003, Th2.2.2.

[32] **Hidayat, A., et al.** Fast Optical Endless Polarization Tracking with LiNbO3 Component. *Proc. OFC/NFOEC'08.* February 24-28, 2008, JWA28.

[33] **Munier, F., et al.** Estimation of Phase Noise for QPSK Modulation over AWGN Channels. *Proc. GigaHertz'03 Symposium.* November 4-5, 2003.

[34] **Wiener, N.** *Extrapolation, Interpolation, and Smoothing of Stationary Time Series.* Cambridge : The MIT Press, 1964.

[35] **Sitch, J.** Implementation Aspects of High-Speed DSP for Transmitter and Receiver Signal Processing. *Proc. IEEE/LEOS Summer Topicals'07.* July 23-25, 2007, Ma4.3.

[36] **Winzer, P. and Gnauck, A.** 112-Gb/s Polarization-Multiplexed 16-QAM on a 25-GHz WDM Grid. *Proc. ECOC'08.* September 21-25, 2008, Th.3.E.5.

[37] **Mori, Y., et al.** Transmission of 40-Gbit/s 16-QAM Signal over 100-km Standard Single-mode Fiber using Digital Coherent Optical Receiver. *Proc. ECOC'08.* September 21-25, 2008, Tu.1.E.4.

[38] **Ip, E. and Kahn, M.** Feedforward Carrier Recovery for Coherent Optical Communications. *IEEE Journal of Lightwave Technology.* September 2007, Vol. 25, 9, pp. 2675-2692.

[39] **Gardner, F.** A BPSK/QPSK Timing-Error Detector for Sampled Receivers. *IEEE Transactions on Communications.* May 1986, Vol. 34, 5, pp. 423-429.

[40] **Godard, D.** Self-recovering equalization and carrier tracking in two-dimensional data communication systems. *IEEE Transactions on Communication.* 1980, Vol. 28, 11, pp. 1867-1875.

[41] **Kikuchi, K.** Polarization-demultiplexing algorithm in the digital coherent receiver. *Proc.*

IEEE/LEOS Summer Topical Meeting 2008. July 21-23, 2008, MC2.2.

[42] **Merziger, G. and Wirth, T.** *Repetitorium der höheren Mathematik.* Hannover, Germany : Binomi Verlag, 1999.

[43] **Noé, R.** synQPSK_PMD_compensation_n02. *Internal report for EIBONE project.* Paderborn : Univ. Paderborn, EIM-E, ONT, 2008.

[44] **Viterbi, A. J. and Viterbi, A. N.** Nonlinear estimation of PSK modulated carrier phase with application to burst digital transmission. *IEEE Transactions on Information Theory.* July 1983, Vol. 29, 4, pp. 591-598.

[45] **Boucheret, M., et al.** A new algorithm for nonlinear estimation of PSK-modulated carrier phase. *Proc. ECSC'93.* November 1993, pp. 155-159.

[46] **Hoffmann, S.** *Hardwareeffiziente Echtzeit-Signalverarbeitung für synchronen QPSK Empfang.* [ed.] University of Paderborn. Paderborn, Germany : Dissertation, 2008.

[47] **Hoffmann, S., et al.** Multiplier-free Realtime Phase Tracking for Coherent QPSK Receivers. *IEEE Photonics Technology Letters.* November 21, 2008.

[48] **Moeneclaey, M. and de Jonghe, G.** ML oriented NDA carrier Synchronization for General Rotationally Symmetric Signal Constellations. *IEEE Transactions on Communications.* August 1994, Vol. 42, 8, pp. 2531-2533.

[49] **Georghiades, C.** Blind carrier Phase Acquisition for QAM constellations. *IEEE Transactions on Communications.* November 1997, Vol. 45, 11, pp. 1477-1486.

[50] **Volder, J.** The CORDIC Trigonometric Computing Technique. *IRE Transactions on Electronic Computers.* 1959, Vol. 8, 3, pp. 330-334.

[51] **Walther, J.** A Unified Algorithm for Elementary Functions. *Proc. Spring Joint Computer Conference.* May 18-20, 1971, p. 379385.

[52] **Noé, R.** Phase Noise-Tolerant Synchronous QPSK/BPSK Baseband-Type Intradyne Receiver Concept with Feedforward Carrier Recovery. *IEEE Journal of Lightwave Technology.* February 2005, Vol. 23, 2, pp. 802-808.

[53] **Cramér, H.** *Mathematical Methods of Statistics.* s.l. : Princeton University Press, 1999.

[54] **Tretter, S.** Estimating the Frequency of a Noisy Sinusoid by Linear Regression. *IEEE Transactions on Information Theory.* November 1985, Vol. 31, 6, pp. 832-835.

[55] **Doerr, C., et al.** Monolithic InP 16-QAM Modulator. *Proc. OFC/NFOEC'08.* February 24-28, 2008, PDP20.

[56] **Noé, R.** Chapter 2: Optical wave propagation. *Lecture notes "Optical Communication.* s.l. : Univ. Paderborn, EIM-E, ONT, 2005.

[57] **Romoth, J.** *Optimierung und Implementierung einer Signalverarbeitungseinheit zur Demodulation von QPSK-Daten.* [ed.] University of Paderborn. Paderborn, Germany : Bachelor-Thesis, 2006.

[58] **Samson, F.** *Entwicklung einer Polarisationskontrolle für die optische Nachrichtenübertragung.* [ed.] University of Paderborn. Paderborn, Germany : Master-Thesis, 2006.

[59] **Wördehoff, C.** *Optimierung einer Polarisationsregelung für die optische Nachrichtenübertragung.* [ed.] University of Paderborn. Paderborn, Germany : Master-Thesis, 2007.

[60] **Cho, P., et al.** Coherent homodyne detection of BPSK signals using time-gated amplification and LiNbO3 optical 90° hybrid. *IEEE Photonics Technology Letters.* July 2004, Vol. 16, 7, pp. 1727-1729.

[61] **Cho, P., et al.** Integrated optical coherent balanced receiver. *Proc. OAA/COTA'06.* June 25-30, 2006, CThB2.

[62] **Ly-Gagnon, D., et al.** Coherent detection of optical quadrature phase-shift keying signals with

carrier phase estimation. *IEEE Journal of Lightwave Technology*. January 2006, Vol. 24, 1, pp. 12-21.

[63] **Delavaux, J., et al.** All-fibre optical hybrid for coherent polarisation diversity receivers. *Electronics Letters*. August 02, 1990, Vol. 26, 16, pp. 1303-1305.

[64] **Grossard, N., et al.** Low chirp QPSK modulator integrated in poled Z-cut LiNbO3 substrate for 2 x MultiGb/s transmission. *Proc. ECOC'07*. September 16-20, 2007, 10.3.5.

[65] **Koch, B., et al.** Optical Endless Polarization Stabilization at 9 krad/s with FPGA-Based Controller. *IEEE Photonics Technology Letters*. 2008, Vol. 20, 12, pp. 961-963.

[66] **Adamczyk, O. and Noé, R.** 13 Gsamples/s 5-bit analogue-to-digital converter for coherent optical QPSK receiver. *Electronics Letters*. 2008, Vol. 44, pp. 895-896.

[67] **Herath, V., et al.** Chipset for a Coherent Polarization-Multiplexed QPSK Receiver. *Proc. OFC/NFOEC'09*. March 22-26, 2009, OThE2.

[68] **Yu, J., et al.** 20x112Gbit/s, 50GHz spaced, PolMux-RZ-QPSK straight-line transmission over 1540km of SSMF employing digital coherent detection and pure EDFA amplification. *Proc. ECOC'08*. 21-25. September 2008, Th.2.A.2.

[69] **Tanimura, T., et al.** Nonlinearity Tolerance of Direct Detection and Coherent Receivers for 43 Gb/s RZ-DQPSK Signals with Co-Propagating 11.1 Gb/s NRZ Signals over NZ-DSF. *Proc. OFC/NFOEC'08*. February 24-28, 2008, OTuM4.

[70] **Duthel, T., et al.** Impairment tolerance of 111Gbit/s POLMUX-RZ-DQPSK using a reduced complexity coherent receiver with a T-spaced equaliser. *Proc. ECOC'07*. September 16-20, 2007, 01.3.2.

[71] **Renaudier, J., et al.** Experimental Analysis of 100Gb/s Coherent PDM-QPSK Long-Haul Transmission under Constraints of Typical Terrestrial Networks. *Proc. ECOC'08*. September 21-25, 2008, Th.2.A.3.

[72] **Pfau, T., et al.** Towards Real-Time Implementation of Coherent Optical Communications. *Proc. OFC/NFOEC'09*. 22-26. March 2009, OThJ4.

[73] **Pfau, T., et al.** Real-time Synchronous QPSK Transmission with Standard DFB Lasers and Digital I&Q Receiver. *Proc. OAA/COTA'06*. June 28-30, 2006, CThC5.

[74] **Pfau, T., et al.** PDL-Tolerant Real-Time Polarization-Multiplexed QPSK Transmission with Digital Coherent Polarization Diversity Receiver. *Proc. IEEE/LEOS Summer Topicals'07*. July 23-25, 2007, Ma3.3.

[75] **Roberts, K.** Electronic Dispersion Compensation Beyond 10 Gb/s. *Proc. IEEE/LEOS Summer Topicals'07*. July 23-25, 2007, Ma2.3.

[76] **Tarighat, A., et al.** Digital Adaptive Phase Noise Reduction in Coherent Optical Links. *IEEE Journal of Lightwave Technology*. March 2006, Vol. 24, 3, pp. 1269-1276.

[77] **Louchet, H., Kuzmin, K. and Richter, A.** Improved DSP algorithm for coherent 16-QAM transmission. *Proc. ECOC'08*. September 21-25, 2008, Tu.1.E.6.

[78] **Seimetz, M.** Laser Linewidth Limitations for Optical Transmission Systems with High-Order Modulation Employing Feed Forward Digital Carrier Phase Estimation. *Proc. OFC/NFOEC'08*. 24-28. February 2008, OTuM2.

[79] **Nelson, L., et al.** Performance of a 46-Gbps Dual-Polarization QPSK Transceiver in a High-PMD Fiber. *Proc. OFC/NFOEC'08*. February 24-28, 2008, PDP9.

[80] **Leven, A., Kaneda, N. and Chen, Y.** A real-time CMA-based 10 Gb/s polarization demultiplexing coherent receiver implemented in an FPGA. *Proc. OFC/NFOEC'08*. February 24-28, 2008, OTuO2.

[81] **Nakashima, H., et al.** Novel Wide-range Frequency Offset Compensator Demonstrated with Real-time Digital Coherent Receiver. *Proc. ECOC'08*. September 21-25, 2008, Mo.3.D.4.

[82] **Pfau, T., et al.** 1.6 Gbit/s Real-Time Synchronous QPSK Transmission with Standard DFB

Lasers. *Proc. ECOC'06.* September 24-28, 2006, Mo4.2.6.

[83] **Leven, A., et al.** Real-time implementation of 4.4 Gbit/s QPSK intradyne receiver using field programmable gate array. *Electronics Letters.* November 23, 2006, Vol. 42, 24, pp. 1421-1422.

10 List of figures & tables

Figure 1.1: Simplified system schematic for the synQPSK project with partners' contributions highlighted 3
Figure 2.1: BPSK (left) and QPSK (right) constellation diagrams 6
Figure 2.2: ASK-8-PSK constellation diagram 7
Figure 2.3: Square QAM constellation diagrams 8
Figure 2.4: Square 16-QAM constellation diagram and bit-to-symbol assignment 9
Figure 2.5: Partial differential encoding for a square 16-QAM constellation 10
Figure 2.6: Optical QAM transmitter structure 11
Figure 2.7: Polarization-multiplexed QAM transmitter 12
Figure 2.8: Polarization mode dispersion emulator (PMDE) 15
Figure 2.9: Polarization diversity coherent receiver frontend 17
Figure 2.10: Coherent optical receiver with analog-to-digital conversion and digital signal processing 19
Figure 2.11: Examples of the phase noise process ψ_k for different values of $\Delta f_{3dB} T_S$ 21
Figure 2.12: Lorentzian carrier power spectra for different values of $\Delta f_{3dB} T_S$ 21
Figure 3.1: Coherent optical receiver structure 23
Figure 3.2: Interfacing between ADCs and DSPU and internal structure of the DSPU .. 24
Figure 3.3: Serial and parallel FIR and IIR filter structures 25
Figure 3.4: Decision-directed carrier recovery with $\Delta = 1$ 27
Figure 3.5: Decision-directed carrier recovery in a realistic receiver with parallel and pipelined signal processing 27
Figure 3.6: Non-data-aided polarization control algorithm 30
Figure 3.7: Decision-directed polarization control algorithm 31
Figure 3.8: Viterbi & Viterbi feed-forward carrier recovery 35
Figure 3.9: Weighted Viterbi & Viterbi feed-forward carrier recovery 36
Figure 3.10: Feed-forward carrier recovery for square QAM constellations 40
Figure 4.1: QPSK carrier phase estimator efficiency for OSNR = 10 dB (left) and OSNR = 16 dB (right) 49
Figure 4.2: Carrier phase estimator mean squared error for different values of $\Delta f \cdot T_S$ 50
Figure 4.3: OSNR vs. BER for Viterbi & Viterbi carrier recovery and different linewidth-times-symbol-duration products $\Delta f \cdot T_S$. 52
Figure 4.4: OSNR vs. BER for weighted Viterbi & Viterbi carrier recovery and different linewidth-times-symbol-duration products. 53
Figure 4.5: Sensitivity penalties at BER = 10^{-3} against $\Delta f \cdot T_S$ for unweighted (a) and weighted (b) Viterbi & Viterbi carrier recovery 53
Figure 4.6: OSNR vs. BER for (S)MLPA carrier recovery and different linewidth-times-symbol-duration products 54
Figure 4.7: Sensitivity penalties at BER = 10^{-3} against $\Delta f \cdot T_S$ for (S)MLPA carrier recovery 55
Figure 4.8: OSNR vs. BER for 4-QAM carrier recovery and different linewidth-times-symbol-duration products 55
Figure 4.9: Sensitivity penalties at BER = 10^{-3} against $\Delta f \cdot T_S$ for 4-QAM carrier recovery 56
Figure 4.10: Phase noise tolerance for different carrier recovery algorithms using either the data from a single-polarization or from both polarizations 57

Figure 4.11: Sensitivity penalty vs. analog-to-digital converter resolution for different QPSK carrier recovery algorithms..................58
Figure 4.12: Sensitivity penalty vs. phase resolution for different QPSK carrier recovery algorithms 58
Figure 4.13: Sensitivity penalty for different numbers of test phase values φ_b for 4-QAM and 16-QAM 60
Figure 4.14: Sensitivity penalty for different numbers of test phase values φ_b for 64-QAM and 256-QAM 60
Figure 4.15: Phase estimator mean squared error and efficiency $e(N_{CR})$ vs. filter half width N_{CR} for square 4-QAM (left) and square 16-QAM (right) constellations with $\log_2\{B\} = 5$ 61
Figure 4.16: Phase estimator mean squared error and efficiency $e(N_{CR})$ vs. filter half width N_{CR} for square 64-QAM (left) and square 256-QAM (right) constellations with $\log_2\{B\} = 6$ 62
Figure 4.17: Phase estimator mean squared error (left) and efficiency (right) vs. filter half width N_{CR} for a square 16-QAM constellation and $\log_2\{B\} = 6$ 63
Figure 4.18: 4-QAM constellation diagram at the receiver before and after carrier recovery for $\Delta f \cdot T_S = 4 \cdot 10^{-4}$ ($\log_2\{B\} = 5$, $N_{CR} = 9$) 64
Figure 4.19: 16-QAM constellation diagram at the receiver before and after carrier recovery for $\Delta f \cdot T_S = 1.4 \cdot 10^{-4}$ ($\log_2\{B\} = 5$, $N_{CR} = 9$) 64
Figure 4.20: 64-QAM constellation diagram at the receiver before and after carrier recovery for $\Delta f \cdot T_S = 4 \cdot 10^{-5}$ ($\log_2\{B\} = 6$, $N_{CR} = 9$) 65
Figure 4.21: 256-QAM constellation diagram at the receiver before and after carrier recovery for $\Delta f \cdot T_S = 8 \cdot 10^{-6}$ ($\log_2\{B\} = 6$, $N_{CR} = 9$) 65
Figure 4.22: Squared distance sum s_b of the b-th parallel block (with test carrier phase angle φ_b) for different filter half widths and different square QAM constellations 66
Figure 4.23: Receiver tolerance against phase noise for different square QAM constellations 67
Figure 4.24: Impact of different linewidth-times-symbol-duration products on the receiver sensitivity of coherent QAM receivers 68
Figure 4.25: Receiver sensitivity penalty vs. analog-to-digital converter resolution for different square QAM constellations 69
Figure 4.26: Receiver sensitivity penalty vs. internal resolution of the distances $\mathrm{Re}[d_{k,b}]$ and $\mathrm{Im}[d_{k,b}]$ for different square QAM constellations 70
Figure 4.27: Receiver sensitivity penalty vs. internal resolution of the squared distance $|d_{k,b}|^2$ for different square QAM constellations 71
Figure 4.28: Influence of the polarization control gain g on the receiver sensitivity 72
Figure 4.29: Start-up development of the matrix elements of the polarization control matrix $\mathbf{M}$ 73
Figure 4.30: Sensitivity penalty vs. normalized polarization control time constant for the non-data-aided and decision-directed polarization control algorithms 73
Figure 4.31: Sensitivity penalty at BER = 10^{-3} for different control gains of the polarization control/ISI compensation algorithm and different values of the FIR filter half width 74
Figure 4.32: BER vs. OSNR for the simplified and original ISI compensation algorithm with $g = 2^{-10}$ and different FIR filter half widths 75
Figure 4.33: Input referred DGD profiles for the emulation of PMD 76
Figure 4.34: Receiver sensitivity for the original ISI compensation algorithm with different FIR filter half widths for different DGD profiles 77
Figure 4.35: Receiver sensitivity for the simplified ISI compensation algorithm with different FIR filter half widths for different DGD profiles 78
Figure 4.36: Developing of $|\det\{\mathbf{M}_i\}|$ for the simplified ISI compensation algorithm with different FIR filter half widths for different DGD profiles 80

10. List of figures & tables

Figure 4.37: Receiver sensitivity (top row) and developing of $|\det\{\mathbf{M}_i\}|$ (bottom row) for different DGD profiles and different FIR filter half widths for the simplified algorithm with optimized start-up sequence. ... 81

Figure 5.1: 800 Mb/s single-polarization synchronous QPSK transmission setup with real-time digital coherent receiver ... 84

Figure 5.2: Single-polarization QPSK transmitter with Bookham optical QPSK modulator. ... 84

Figure 5.3: LiNbO3 optical 90° hybrid and associated control unit ... 85

Figure 5.4: Differential photodiode pairs and amplifiers ... 85

Figure 5.5: Commercial ADC board (left) and FPGA board (right) for digital signal processing ... 86

Figure 5.6: BER vs. preamplifier input power for synchronous QPSK transmission with self-homodyne detection using an ECL. ... 87

Figure 5.7: BER vs. preamplifier input power for synchronous QPSK transmission with self-homodyne detection using a DFB laser. ... 87

Figure 5.8: Laser beating at the output of the coherent receiver frontend ... 88

Figure 5.9: BER vs. preamplifier input power for synchronous QPSK transmission with intradyne detection using DFB lasers. ... 88

Figure 5.10: Measured BER vs. optical preamplifier input power at 1.6 Gb/s data rate. ... 89

Figure 5.11: Transformation of the 3x3 coupler outputs to I&Q signals ... 90

Figure 5.12: Synchronous QPSK transmission setup with clock recovery and symmetrical 3x3 coupler ... 91

Figure 5.13: Receiver sensitivity of a synchronous QPSK receiver with either a symmetric 3x3 coupler or a 90° hybrid. ... 91

Figure 5.14: Simulated and measured BER floors for different linewidth-times-symbol-duration products ... 92

Figure 5.15: 2.8 Gb/s polarization-multiplexed QPSK transmission setup with a real-time FPGA-based synchronous coherent digital I&Q receiver ... 93

Figure 5.16: Block diagram for the pattern generator and QPSK precoder implemented in an FPGA ... 93

Figure 5.17: MK325 Xilinx Virtex-II ProX RocketIO characterization board used for pattern generation and precoding ... 94

Figure 5.18: Polarization-multiplexed QPSK modulator. ... 95

Figure 5.19: Polarization scrambler and the corresponding fast polarization changes displayed on the Poincaré sphere. ... 96

Figure 5.20: Polarization diversity coherent optical receiver frontend. ... 97

Figure 5.21: Commercially available analog-to-digital converters (left) and an FPGA board for digital signal processing (center). ... 97

Figure 5.22: Influence of the carrier recovery filter width on the receiver sensitivity and BER floor. ... 98

Figure 5.23: Measured BER vs. optical preamplifier input power at the coherent receiver for the two different control time constants of a) 12 µs and b) 3 µs and various polarization change speeds. ... 99

Figure 5.24: Receiver sensitivity penalty vs. scrambling speed for a bit error rate of 10^{-4} ... 100

Figure 5.25: Long-term BER measurements with polarization change speeds causing 1 dB loss in receiver sensitivity. ... 100

Figure 5.26: BER (I&Q averaged) vs. optical power at the preamplifier input for different speeds of polarization change. (a) for c = 3.0 µs (slow control) and (b) for c = 0.75 µs (fast control). ... 101

Figure 5.27: Receiver sensitivity penalty vs. speed of polarization changes on the Poincaré sphere. ... 102

Figure 5.28: BER vs. optical preamplifier input power for 0 dB and 6 dB of PDL. ... 102

Figure 5.29: SiGe ADCs and CMOS chip package integrated in the synchronous QPSK testbed. 103
Figure 5.30: Bit error rate vs. preamplifier input power for synchronous polarization-multiplexed QPSK transmission at 10 Gb/s for different polarization states at the receiver input. 104
Figure 5.31: Influence of the filter half width N_{CR} in the carrier recovery circuit on the BER performance of a 10 Gb/s polarization-multiplexed synchronous QPSK transmission system 105
Figure 5.32: Comparison of 5 Gb/s single-polarization QPSK transmission against 10 Gb/s polarization-multiplexed QPSK transmission 106
Figure 6.1: Milestones in synchronous QPSK transmission with real-time coherent digital receivers 107

Table 2.1: Differential coding penalty for different square QAM constellations 11
Table 3.1: Barycenter carrier recovery filter structures 38
Table 3.2: Required hardware components for different carrier recovery algorithms 43
Table 4.1: Maximum tolerable linewidth for 10 Gbaud systems with different square QAM constellations 67
Table 4.2: Required symbol rate and maximum tolerable linewidth to realize a 100GbE (112 Gb/s) system with different square QAM constellations 68
Table 4.3: Analog-to-digital converter requirements for a polarization-multiplexed QAM transmission system for 100GbE 70

Acknowledgement

First of all I want to thank Prof. Dr.-Ing. Reinhold Noé for giving me the opportunity to do research on this interesting topic of real-time coherent optical transmission. To work under his supervision was both challenging and exciting. Many fruitful discussions and valuable suggestions have substantially contributed to the success of this work. Moreover I would like to thank him for his confidence to send me to several conferences as his representative. The gained experiences and established contacts are invaluable.

I also cordially thank Prof. Dr.-Ing. Ulrich Rückert for the trustful cooperation within the synQPSK project and for the assessment of my dissertation.

Special thanks go to all my colleagues in the working groups "Optical Communication and High Frequency Engineering" as well as "System and Circuit Technology". Dr.-Ing. Sebastian Hoffmann was a great officemate and was willing to discuss even the most wacky ideas. Dr.-Ing. Suhas Bhandare was an invaluable source of help for any kind of problem in the laboratory. Dr. Olaf Adamczyk was a great advisor and significantly improved the quality of my publications. Dipl.-Wirt.-Ing. Ralf Peveling and Dipl.-Ing. Christian Wördehoff were indispensable for the FPGA implementation of the algorithms and additionally gave substantial moral support.

I would further like to thank all partners of the synQPSK project. All my experimental work would have been impossible without the components provided by them. In this context also the funding by the European Commission should be mentioned. Its continuing support and patience in spite of obstacles and delays was not self-evident.

I am also grateful that the International Graduate School "Dynamic Intelligent Systems" granted me a scholarship. Next to the skills and knowledge I gained in different lectures and workshops they offered me the opportunity to build up a network of friends that spans all around the world.

Heartfelt thanks go to my family. My whole life they gave me absolute love and support.

Finally I want to thank Luz-marina Guitard for her love and passion that help me to become the person I want to be.

Die VDM Verlagsservicegesellschaft sucht für wissenschaftliche Verlage abgeschlossene und herausragende

Dissertationen, Habilitationen, Diplomarbeiten, Master Theses, Magisterarbeiten usw.

für die kostenlose Publikation als Fachbuch.

Sie verfügen über eine Arbeit, die hohen inhaltlichen und formalen Ansprüchen genügt, und haben Interesse an einer honorarvergüteten Publikation?

Dann senden Sie bitte erste Informationen über sich und Ihre Arbeit per Email an *info@vdm-vsg.de*.

Sie erhalten kurzfristig unser Feedback!

VDM Verlagsservicegesellschaft mbH
Dudweiler Landstr. 99   Telefon +49 681 3720 174
D - 66123 Saarbrücken   Fax   +49 681 3720 1749

www.vdm-vsg.de

Die VDM Verlagsservicegesellschaft mbH vertritt

Printed by Books on Demand GmbH, Norderstedt / Germany